THE EXCYLES

THE TRUE EXPERIENCES OF A WOMAN WHO
IS LOVED BY EXTRATERRESTRIALS
AND WAS ROMANCED BY A
U.S. GOVERNMENT INTELLIGENCE AGENT

BY MIA ADAMS

EXCELTA PUBLISHING, INC.
FORT LAUDERDALE, FLORIDA

Excelta Publishing, Inc.
P.O. Box 4530
Ft. Lauderdale, Florida 33338

Printed in the USA - 1995
Second Printing - 1/96

ISBN# 0-9649905-0-4
Library of Congress Number Pending

THIS BOOK IS DEDICATED TO THESE SPECIAL SOULS

MY MOTHER, IDA

THE LATE LEONARD STRINGFIELD

ZARG AND EXCELTA

THE SOUTH FLORIDA AERIAL PHENOMENA INVESTIGATORY GROUP

THE EXCYLES

MUCH THANKS, LOVE AND LIGHT.
MIA ADAMS

CONTENTS

CONTENTS *(continued)*

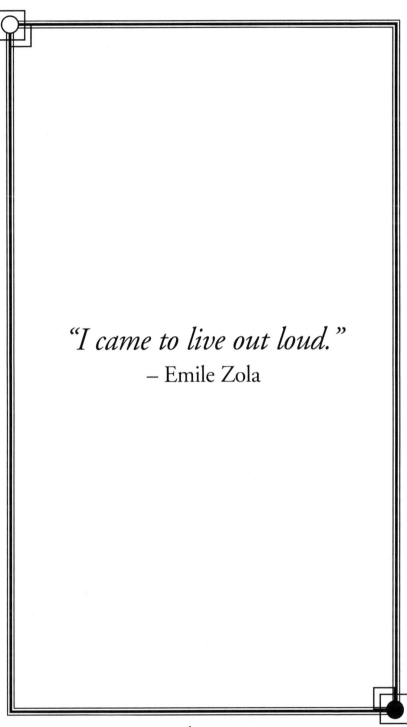

"I came to live out loud."
– Emile Zola

WHY THIS BOOK?

Among the things I covet most in my life are my freedom and my privacy. In writing this book, I am exercising my freedom of expression at the great risk of my privacy. This is a chance that I feel I must take, for in many ways it seems I have been predestined to do this. I have been reluctant to chronicle my experiences and the information I have been given, due to the incredibly unbelievable nature of them. Admittedly, if I hadn't experienced them myself, I probably would have shook my head and rolled my eyes heavenward thinking that Mia Adams' nose would eventually give Pinnochio's a run for his money.

All of what I have written are my true experiences. I have found these experiences powerful and important. Others that have shared information or these experiences with me have repeatedly suggested, quite strongly, I might add, that I write them down as a book. After my experiences with an Intelligence Agent of the United States Government, I, too, felt this had to be done. Where there is information that I have been given by others I have documented this clearly as *their* information and not mine. I have tried to corroborate data where I could but it was not always possible to do so. These are times where things we hear, see and read are not always what they seem but I have tried to stay focused on being objective and rational, although there are many experiences in my life that have gone way beyond the pale of mundane reasonableness. I have tried my best to ferret out logical explanations but where I couldn't, I let my gut instinct and woman's intuition lead me to the appropriate assumptions. Sometimes it made me take that giant leap into the spiritual realms of New Age and New Thought and most unspiritually, New Government Paranoia. Many times I did this kicking and screaming mentally but I am afraid, dear Reader, this had to be done. Especially in the area of paranoia, perfect paranoia became perfect awareness! For all intents and purposes, this has become the New Logic for me and I think it is a good one! The combination of factors of spiritual other-worldliness and heavy handed government intrusion has made my life quite paradoxical, if not downright magically mysterious!

My friends say that if people would know how centered and normal I am, they would find what has happened to me more credible. But since you all don't know me, as an apologia, I will tell you that I am very sane, indeed. The true test of my sanity is that I have been able to deal with what has been offered up to me as tests and have weathered the most wacky and perhaps even potentially dangerous experiences and still find my life an incredible adventure.

The only things that are not true are the names of most of the people and some of the places I have written about. My life has certainly been stranger than fiction and at the same time quite miraculous. The best thing about life is that we never completely know the possibility or the meaning of the next moment.

Emile Zola, the French novelist, said, "I came to live out loud!" If Emile was here, I would say to him, "Me too, Emile!"

"THE ACCOUNT I AM ABOUT TO GIVE IS BASED ON MY EXPERIENCES AND THE PERSONAL INVESTIGATION I HAVE CONDUCTED OVER THE PAST SEVERAL MONTHS. I CANNOT, AND IN A FEW INSTANCES I AM SPECIFICALLY PREVENTED FROM, REVEALING WHERE AND WHEN I ACQUIRED ALL OF THE INFORMETION I WILL PRESENT. I WILL STATE THAT ON TWO OCCASIONS I DELIBERATELY VIOLATED SECURE AREAS AND ON ONE OCCASION PARTICIPATED IN A RATHER UNORTHODOX FIELD INTERROGATION IN ORDER TO OBTAIN DATA. I ALSO DELIBERATELY INVOLVD SEVERAL INNOCENT PARTIES WHO, UNFORTUNATELY, MAY BECOME SUBJECT TO PENALTIES WHICH SHOULD RIGHTLY OnLY BE MINE. I HAVE LIED AND I HAVE DONE SO KNOWINGLY AND REPEATEDLY IN ORDER TO ELICIT INFORMETION AND IN ORDER TO CAUSE GOVERNMENTAL ENTITIES TO REACT TO MOVES THEY OnLY THOUGHT I WAS MAKING. I FREELY ADMIT THAT I AM GUILTY OF ABUSING THE POWER OF MY OFFICE AND MY CREDENTIALS. I DO NOT APOLOGIZE FOR THIS, BUT I MAY SOON BECOME SUBJECT TO THE ADMINISTRATIVE OR LEGAL CONSEQUENCES OF THESE ACTIONS. SO BE IT. AT THE RISK OF APPEARING MELODRAMATIC, MORE DIRE CONSEQUENCES MAY ALSO FOLLOW. I CAN OnLY STATE THAT, IF RESISTANCE IS WITHIN MY POWER, I WILL NOT GO QUIETLY..."

-Opening paragraph, reproduced verbatim from the actual "unofficial" report that was hand-delivered to Mia Adams by a verified legitimate FBI agent. The entire report appears on pages 127-139 The misspellings and "typos" remain a mystery, as this agent was highly educated and otherwise showed great proficiency in spelling and vocabulary.

CHAPTER ONE

Big Brother & Me

"If you ain't the lead dog, the scenery never changes."

– Old Alaskan Saying

PANDA UNCAGED

My life truly is a strange and wondrous journey. I always knew that I had a book in me, but I never expected that it would end up to be one that read like a John LeCarre spy novel! The book that I started to write did not end up being the one that I have now written. Here is how it began.

In August, 1988, I became aware that I have been in contact with extraterrestrial and or interdimensional intelligence all my life but was not allowed to be fully conscious of it. I was so in awe and wonderment at this revelation that it started me on a new path of exploration. I had been a metaphysical seeker for the prior ten years or so but my interests did not lie in space. I never doubted that intelligent life exists beyond the Earthly realm, but it was not a focal point in my metaphysical pursuits. I had no information as to concrete evidence one way or other about these beings and even though I was sure of my own experience, I was confused as to the facts. It was so baffling to me that I decided to see if I was the only nutcase to have this kind of experience. The only others I can recall hearing about were those whose lives were exploited in the tabloid headlines I read as I was waiting in the supermarket checkout lines.

I decided that the best way to pursue this quest for information was to put a small ad in the classified section of a well known magazine for the metaphysically oriented. The ad read as follows:

EXTRATERRESTRIAL CONTACT RESEARCH.
Have you had direct, telepathic or other contact with E.T.s?
Write me about your experiences. Confidentiality assured.
Please include your name, address and phone number.
Atena Adams ———

I did not know what to expect and I got much more than I ever anticipated! That is true both in numbers of replies and in quality. I took out a P.O. box because I didn't want a bunch of potential

3

schizos knowing where I lived. I also used my "space name" instead of my real one to cover my anonymity in all areas. It also seemed more fitting since it was supposed to be my name from the last incarnation I had from Excelta (at least that is what my space contacts told me.) I always felt comfortable with Atena, even though I liked my given Earth name, Mia.

I must say that amongst the 200 or so replies, I did not receive any that merited the whacko bin. They all were quite sincere. Whether or not they all emanated from a true source of experience, I cannot judge. I was always happily anticipating my trek to the post office and for months they yielded much in the way of confirmation of my experiences and amazement at the commonality of others experiences, as well. But, there were two replies that were quite different from anything that I expected or even knew about!

The first one came in October, 1989. I knew that it was going to be a real strange one because of the way that the envelope was. It had no return address but had a California postmark. It was sealed with a kind of a cellophane tape I had never felt before. Even though it was transparent, it was impossible to pull away from the paper of the envelope, and it could not be torn. Through the mid point of the seal a straight line was drawn vertically. Even with my untrained detective's mind, I surmised that was supposedly a deterrent to tampering by an unauthorized reader without the recipient's knowledge. "Good God," I thought, "what could be inside this that could have merited such concern?"

The innards of that epistle to me were really something out of the UFO lore I had just begun to learn about, and it was of the most spectacular kind.

The writer who wanted to be called by the code name "Panda" claimed to be connected with the clandestine, super-secret happenings at Nellis Air Force Base. Panda's job was to organize and schedule movement of personnel and visitors, with high security classifications, from Edwards Air Force Base in California to Nellis. He stated that he had worked at Edwards for several years and knew of "Dreamland" operations (a super secret site at Nellis) and daily CIA operations covered by Lockheed, the aviation corporate giant. Panda said that his information goes way beyond most people's comprehension and *without a doubt could cost us both our lives!* He continued that if I really wanted to know this stuff, I had to be careful because people often come up "missing."

Nellis Air Force Base, near Las Vegas, Nevada, is well known to ufologists for the bizarre scenarios that are supposed to ensue there. Apparently, according to many documented reports, space aliens and the U.S. government have special areas of the base called Area 51, the Skunk Works, Groom Lake and other specific points within, where there are nefarious activities of all kinds. Particularly macabre are reports of biogenetic experiments with humans, animals and hybrids of many types. It is also said that there are alien spacecraft and technology that are available to humans for research and use. The use of these facilities are said to be synergistic for the aliens and humans.

Panda was interested in only a very covert manner of communication with me since he stated the extremely dangerous nature of his revelations. We were to communicate via a specific coded reply with phone conversations delivered from him to me very early in the morning. Since we were three time zones apart, I would be snoozing deeply in Morpheus' arms by the time he was ready to divulge more of this terrifying and sinister information.

I was truly perplexed and frightened. In my quest for information for my personal knowledge about ETs and UFOs, I was ready to go the distance but I was not prepared to tempt death's door to get it! However, the metaphysician in me spoke of Fate's presenting me with something that should be dealt with, otherwise why would it have come when I did not ask for it! Convoluted logic that it was, I was titillated enough to take the ball and run and see where it lead me. The adrenaline was pumping but I would definitely need some kind of help and counsel.

At that point in time, neophyte ufologist that I was, the only person I could think of and could count on who had any expertise in this subject was my MUFON friend Martin. MUFON (Mutual UFO Network) is a UFO organization that investigates and discusses current and past UFO phenomenon of all kinds. It is a more nuts and bolts organization than some where spiritual and metaphysical approaches are deemed more important to the ufological experience than the third dimensional ship that got the space beings here. I went over to Martin's house and we discussed the possibility that this could really be a great opportunity for investigation. Martin knew of all the allegations about Area 51 and was savoring the chance at getting some unique information.

A man named Robert Lazar, at that time, was also revealing some truths about Nellis. He stated that he had personal, first-hand experience as a scientist working on "Above Top Secret" projects, so the letter writer's revelations were not new but perhaps he could shed some additional light on the subject.

I was to write back to Panda. I included Martin's phone number and told Panda that Martin would field the calls since I was too timid at that time to deal with that kind of reality. Several phone calls did manifest in the very early morning hours. Martin was so groggy, he could hardly write legible notes, but it did seem to him that Panda, indeed, was a valid employee of an allied agency that dealt with such matters. We received some nifty information about the pursuits and aims of these projects, some we were already aware of and others that were mind boggling.

Panda was 37 years old and an employee of the Department of Energy for five years, three of those years at the Dreamland facility at Nellis. His father had also been employed by the federal government, working on special projects in the 1950's as an

6

aeronautical engineer. The business of government was nothing new to Panda. His interest in things not of this world catalyzed when he had a conscious abduction by aliens at the age of nine. As a young married couple living near Cape Canaveral, he and his wife and their dog had two conscious encounters with aliens and sightings of a humming, shoebox shaped craft. In doing investigation about their experiences, his interest in the extraterrestrial hypothesis became very intense. It, therefore, was not difficult for him to garner enthusiasm to be employed in his very special kind of work. Soon into the game, he became disillusioned at what he saw and heard. He felt that sometime in the near future the existing government would collapse and the new government would not represent the true interests of its citizens; hence, his desire to be a "Cosmic Deep Throat."

Panda had much information on the infrastructure of the conspiracy of the United States of America's working with extraterrestrials. He said that many government agencies were involved but the employees within those agencies were given security clearances and committed to oaths that were well above top secret. It was very dangerous to breech those oaths or even just to appear to be a security risk. The governmental agencies involved, according to Panda, were the Defense Advance Research Projects Administration (DARPA), Department of Energy (DOE), Central Intelligence Agency (CIA), Defense Intelligence Agency (DIA), U.S. Navy, Air Force and Army. The money to run this program came from the "Black Budget" and was filtered through DARPA. The "Black Jack" program was for the recovery and retrievals of extraterrestrial space craft, hardware or crashed materiel. It also was the name given for the propaganda and disinformation machine connected to the project and for the aerial surveillance facility network for sightings and radar contacts of UFOs. Panda related that the Majestic Project or MJ-12 was real and provided the infrastructure and compartmentalization for the coverups. MJ-12 is said to be historically the main operation for dealing with UFOs, sightings and disinformation related to the truth of the subject. He also mentioned a lawless group connected to the Department of the

Navy (DON). He said that the Navy has been manipulated via the intelligence community and has become the shell operation for the entire conspiracy. The infamous "Black Helicopters" are flown by the CIA and the Drug Enforcement Agency (DEA). Panda was not aware of any other agencies using these.

There are reportedly several underground facilities across the United States that are used to deal with retrieved space craft and are human/alien workplaces where laboratories and bases are used in common. Panda was most familiar with the underground facilities at Edwards Air Force Base, California; Las Crusas, New Mexico; Cheyenne Mountain Norad Complex near Colorado Springs, Colorado; and Dreamland ("The Ranch") at Nellis Air Force Base, Nevada. Very much involved strategically, also, were the China Lake Naval Air Station, Scott Air Force Base and Homestead Air Force Base. Panda stated emphatically that the reputed Roswell crash UFO was stored and hidden at Edwards Air Force Base. There have been other discs there that were too big to be shipped out, so they had to be flown out. Panda didn't know whether it was aliens or humans or both who did that. There were also private defense and aerospace contractors in Southern California that had underground facilities with very strange things going on. Panda said that Northrup had an enormous underground facility near the Tehachapi Mountains, with over 40 levels below ground. Northeast of Llano, California, McDonnell Douglas Company operated a facility with very odd runways where glowing objects have been seen at night. He stated that Lockheed and General Electric also had underground facilities with high tech scenarios that have been reported connected to covert military and alien activity. There was no doubt in Panda's mind that these were all connected with the conspiracy.

The Cosmic Deep Throat stated that there were nine discs on site at Nellis at the time, and also live aliens working with a large scientific contingent there. One group of aliens that was incarcerated and died there was called "Ashue". They were subjected to experiments and a Faraday Cage. They were both telepathic and vocal. He claimed there were several groups of aliens involved besides the well-known Grays. The "Blues" had similar features to the Grays and were three and one half to four feet tall. The "Whites" were lighter in color and had smaller eyes. There were also some Grays who had darker facial features and appeared like raccoons. Panda's opinion was that the Grays had little regard for life. They were said to be violent, deceitful and totally results-oriented, like our scientists who were involved with the aliens. He said that we also have been in contact with the Orion group and "Venusians." The Pleidians have avoided contact with governments. According to Panda, the human/alien project has been going on for more than 30 years, and the Nellis/Dreamland base has participated since 1977.

Panda explained that not all of the scientists involved had "Dr. Strangelove" mentalities. Many researchers would lay awake nights and became emotionally unstable because of what they were seeing and participating in. Paranoia was running rampant, since the motives of these morally-concerned people were now suspect. With knowledge came responsibility. They were now questioning the moral fiber of those in authority, but once they were involved, they could not ever get out of the program. Panda wanted out, but knew that it was most likely impossible.

Most of the veterinarians were on three to four month contracts due to the horrors of their work. Some never reported back. It was too emotionally burdensome. As reported by abductees who have seen them, live animals *did* have plexiglass panels inserted in their bodies to view the organs at work. Animal mutilations were justified as being "scientific investigations." Panda felt that the scientific exchange of data has been lopsided, with the alien contingent getting the greater amount and quality of information. Dialogue was said to be perplexing to aliens. Sharing was not their natural proclivity. Some of the technology exchanged was written in what the cognoscenti call "The Yellow Book." It was written as a thesaurus related to information applied to our practical sciences, navigation, electromagnetic propulsion and particle beams.

"Apparently, aliens know how to 'fold four navigational points' for interspace travel. They can also do this through electromagnetic flux and reflection. The Bermuda triangle is an entry point for most alien visitors. They report that they use the Pyramids as a focus of electromagnetic reflection," Panda stated.

He went on to say that our physicists working with the aliens were working under the guise of the Department of Energy. They did not understand the physics of the "four point fold" yet, but many continuously worked on it, on rotations of 24-hour shifts. It was reported that we used particle beam weapons to shoot down a disc over South Africa. Apparently, the disc was not one from the aliens we were in cahoots with. The motive of alien operations here was human betrayal. Panda said that human history would be rewritten by the government with the aid of this particular alien confederation.

Panda related a topic equally sinister, concerning our involvement with Nazis. During the decoding and dismantling of the German war machine after the Second World War, glyphs were uncovered that told of the history of how this planet was seeded by Pleidians, Grays and Blues. Some of the Nazis were given sanctuary for their knowledge of this and other Nazis were assisted movement to South America by the CIA. Most of the glyphs went to underground storage at Dreamland.

Panda stated that the use of mind control on the masses was also an aim. "The Army uses Neuro Linguistic Programming in psychological warfare and mind control scenarios. Advertising and subliminal programming are now being used for Gray propaganda. We have seen examples of this in television commercials and print media. Intense Gray propaganda is focused on Los Angeles and San Diego areas. Grays will be created as gods for the masses through fear, worship and corruption. Psychological material has been created for this. It is imperative that we seek within ourselves for spiritual sustenance and guidance or else we will lose control!" he insisted.

After a few months, Panda asked to speak with me, if I felt inclined to do so. I had become a bit more courageous and agreed. Unfortunately, Panda's phone call came at a very untimely moment for me. That particular night, I had suffered from insomnia and for the first time, had taken a sleeping pill. Since I waited until the wee hours of the morning to do so, it unfortunately coincided with the time that Panda chose to call me. I was totally incoherent and cannot remember one thing we discussed. I am sure that he thought I was some sort of a moron. I do remember that I was very apologetic but I don't think that sufficed. At any rate, we have not heard from him since. That probably is for the best. In retrospect, I am suspicious about my taking that sleeping pill just before he called. Could I have been under mind control?

Panda, who seemed to be extremely paranoid and took all kinds of precautions about contact with us, might have been spied on and my connection with him was noted and followed up. Or Panda, himself, might have been connected with an intelligence group and through machinations got my attention to see how far I would go and what kind of a researcher I was. At any rate, there is a good chance, that our contact was more than meets the eye or pen.

Over five years later, in late 1994, I would meet a government agent who would validate much of what Panda had stated. This agent also verified personal information Panda had given about himself. Whether this information is true or not, the reader must decide. I, personally, hope it is one huge massive lie or disinformation campaign. Regardless, somehow and in some way, I had been chosen to be a conduit for this information to surface. In my search for the reasons, much adventure transpired and many miracles have evolved from it. As I experienced all of them, I shook my head in amazement. "Why, God, *me?*"

"Corey's intention is to harass and interfere in people's lives. His interest is in disrupting you and making you unsure of yourself. His ultimate aim is for you to question everything you believe."

– Zarg
(Mia Adams' extraterrestrial
contact from Excelta)

THE COREY CONNUNDRUM

In October, 1989, a month after I heard from Panda, I received a letter from Corey Elder. It came with postage due, as it was a very thick mound of 25 handwritten pages. It was quite a complicated bunch of words filled with military jargon, and like Panda's letter, much more than I bargained for.

Corey began his letter stating that he was a black man currently residing in a state penitentiary and he hoped that these two facts would not cloud my perception of what I was about to read. He proceeded to explain in a very articulate fashion facts that he said he had never revealed before. It was my ad in the magazine that caught his attention. He said the magazine was on a prison guard's desk and Corey asked to borrow it.

Prior to my marriage, I had been a Probation Officer and had plenty of experience dealing with the criminal justice system, prisons, prisoners and guards. I don't mean to be a bigot about prison turnkeys, but I find it totally remarkable that a spiritually-oriented magazine would be a favorite of a prison guard, especially in a prison in the Deep South! Something seemed incongruous about this. Such a quirk of fate brought out my cynical side. This guard thing has always been an uncomfortable connection for me. From the beginning, it felt wrong.

Corey sensed that my ad for research might be the perfect avenue to vent the secrets that had created so much fear and pain in his life. He then stated in detail an account of his experiences as a member of the United States Air Force in 1983. In the spring of that year, Corey was a second lieutenant in the Air Police at Mountain Home Air Force Base in Idaho. A surprising event occurred there. In great detail, Corey elaborated on what is known as a "crash retrieval scenario." A UFO was spotted on the radar, followed by a crash in the area, followed by a pro-forma retrieval by the "Blue Beret" retrieval team. The base of operation for the team and the site for the retrieved disc was Mountain Home.

Purportedly Corey and a buddy got involved in this scenario in a fearsome way. While sneaking a cigarette in the Quonset Hut encapsulated disc area they were supposed to secure, curiosity got the best of them, and they peeked at the disc through a crack. What followed for them was a tale of Air Force Hell.

What followed for me was three years of communication with Corey and research into his tale. I tried my best to obtain verification from Corey and from the available records. I had opportunities to speak to Corey on the telephone. I wanted to hear for myself the timbre of his voice, to see if he was telling the truth or lying creatively. Since I have had professional experiences with criminals in my past, I was very aware of convicts' need to communicate with the outside. Many fabricate in their minds scams that they seek to perpetrate, even from the inside of jails and penitentiaries. What was so intriguing about Corey was his fund of correct information about military procedures and UFO data. If he was lying, he was an excellent student and/or had some great informant teachers. He did seem to know what he was talking about.

The most winning play in Corey's game was that he *never* asked for money. That did not seem to be his intention or did he want to put a romantic tone to our investigation. The normal modus operandi for convicts are these two ruses. Corey sent me "his military records," as I requested. He said that he was copying them for me. They appeared to be very complete and in keeping with his story. There even was a reference to his being disciplined for infractions of security while on duty. Very perplexing for me was the fact that the many pages were held together with a metal closing bracket. This metal piece would make a fine weapon for the crafty con. In the prison systems, ostensibly they are very, very careful about anything metallic. I don't know how Corey got that bracket but I don't think it was normal. I decided not to ask him, it might offend him. But I would keep that fact in my mental file, just in case things started to look off-kilter and not valid.

14

Another fact that really fascinated me was Corey's social security number. He was supposed to be, according to his documents, 20 years younger than I am. Yet, his social security number had the first five numbers exactly the same as mine. Numbers on social security cards are given according to the city and the year that you applied for it. How did Corey get a number so close to mine? I asked him where he got his card. At first he said that he didn't remember since his mother applied for it. Eventually, he remarked it might have been Chicago, since he lived there when he was in high school. He also told me he had lived in Ft. Lauderdale and went to high school here. The name of the high school was partially incorrect. It sounded to me as if someone gave him the name to tell me and he made an error. I still could not fathom how he could have a number so close to mine. How synchronistic for Corey to have lived in two cities that I have. That seemed very mysterious or very conveniently orchestrated.

Corey said that he was in jail pending a trial for a murder he didn't commit. He said that he had been followed by a government type in a dark suit while he was walking in a deserted area at night. Corey panicked and shot him with a gun he carried because he was afraid of the bad government men who knew he had seen a space craft at Mountain Home Air Force Base. There was a lot of soap opera related to his story, but Corey was so soft spoken, polite and well-versed on his subject, it was difficult to discount him completely.

Within the first year of our correspondence, I received several unsigned letters that were of an intimidating and diabolical tone from supposed intelligence agents who knew that I was in communication with "Lieutenant Elder" and wanted to know what I was going to do with the information I had. Postmarks showed that they were from the Lancaster, California area, (like Panda) and from Las Vegas, Nevada. Interestingly enough, Lancaster is near Edwards Air Force Base and Las Vegas is near Nellis Air Force Base, both sites of UFO intrigue. If Corey was having letters sent from there to intimidate or convince me, he was pretty smart to do it. I

15

wasn't intimidated, nor was I convinced, I just wasn't going to throw the baby out with the bathwater! At least, not yet.

I had not been able to corroborate the crash of the disc in 1983 or anything special happening at Mountain Home, in general. Even local UFO researchers in Idaho had not heard rumors of such a crash. I decided to probe Corey's crime and asked to contact his attorney. He gave me an implausible story about an attorney he had who was in Texas (not the state Corey was incarcerated in). The attorney had been suggested by the prison chaplain, since the attorney was well versed in military related crimes. This also seemed weird. I asked if I could contact his attorney to substantiate his problems. Corey said he would have his attorney contact me.

Within a week, I received a letter from "his attorney." Just looking at the envelope from the attorney made me suspicious. Attorneys, in general, have good quality, business-like stationery, printed with their addresses and phone numbers on the paper. The envelope had a printed sticker with the attorneys' partnership names on the envelope. My address was typed on a sticker and placed on the envelope. I had never seen that done before. It was very unprofessional and cheap looking. The stationery, itself, had a letterhead with the names of the attorney partners, address but *no phone number!* What kind of legal partnership doesn't have a phone? Phony ones! I contacted the Bar Association in the city the attorney supposedly practiced in. There was no record of him or his associates. I contacted the state registrar for licenses and they had no licenses to practice or records of passing the bar for this "attorney". Obviously, this was a hoax. The letter was postmarked from Texas, so the address was probably a mail drop!

What was this all about? Where was this going? There seemed to be a concerted effort to make me believe this story, but to what end purpose? When I spoke with Corey, I told him of my concerns about his attorney. He feigned surprise and said he thought his attorney was on the up and up, since the chaplain had recommended him. Corey showed concern that *he* might be getting conned! Corey was really smooth and good at his game, whatever it was. He wanted to show his gratitude to me for being such a good friend to him by giving me a television! He said that it was the only thing he had of value and wanted to send it on to me.

Now, for the first time since I got involved with Corey, I was getting concerned. I doubted if he had his own television set in prison and, even if he did, there was absolutely no way he could send it to me. I don't think United Parcel Service makes regular pickups at prisons. Perhaps some kind of contraband or explosives were meant to be enclosed in the television set. I thanked Corey and said no. He insisted that I accept his gift and kept on doing so for several weeks, in letters and verbally over the phone. I told him that he better not send me anything like that because I would not accept it. He finally got the message and stopped asking.

Soon after Corey gave up on the television scenario, I received a phone call from Charles Carter. He said that he was an acquaintance of Corey's. I did not like this at all. Why and how did Carter get my phone number? No way was this supposed to happen. Corey had promised that my phone number would be committed to his memory. Carter said that Corey had given him my number when they were in prison together. Carter wanted me to write to Corey and give Corey Carter's telephone number. He said that it was easier for him to call me long distance than it was for him to write to Corey. The story sounded very fishy to me and I was uneasy.

Carter then proceeded to tell me why he wanted Corey to get his phone number. He said that he was supposed to "help Corey get out of the situation that he was in." Carter then told me he had been in Special Forces and had a lot of abilities to help get Corey out of where he was. I immediately interrupted Carter by telling him that I had a notion of what he was going to tell me and that I did *not* want to hear it or have his phone number. I only wanted to do things legally and I believed that Corey should use the system and due process. I reiterated that I would *never* do anything illegal! He said that he guessed that he should hang up then. I agreed that would be best.

He did not. He went straight back to the immediate conversation. He said that he wasn't interested in helping Corey out because he liked him, but he would do it for money, and that Corey had a lot of money. There was nothing about Corey that Carter spoke well of. It was a real Corey-bashing session. The only thing that Corey had in common with Carter was that they both were Masons and Shriners. Now that was a really strange correlation! Quite suddenly, Carter ended the conversation by saying that he'd better hang up. He didn't even say goodbye. I put the phone back in its cradle, shaking my head in confusion. Now what was that conversation supposed to infer?

Perhaps it was some kind of test for me to see if I would do anything illegal. I don't know why anyone, including Corey, would ever think I would. There also was a lot of disparagement of Corey and sly remarks about my not really knowing who or what Corey was. The super-negativity appeared to be overkill. But again, why? Why would he have a friend call to discredit him like that? I had not promised Corey anything, except to listen to his tall tales. Nothing made sense.

I was anxious to speak with Corey about his jailbird buddy. I had a feeling that Corey would call soon, perhaps to test the waters or smooth things over. The next morning I received a call from the Sprint operator asking if I would accept a call from Corey. When I accepted, the phone went dead. If it was Corey, he did not phone back. I thought this was also really strange. Either he did not really call me, or if he did, perhaps he changed his mind about what he wanted to say, or else he just plain chickened out! When I was finally able to contact Corey, he vehemently denied that he knew Charles Carter, or that he had phoned me and hung up, or that he gave my name or phone number to anyone. Maybe this was a prison personal vendetta, but why try to get me involved with nasty, illegal things? I never found out the answer to that.

Concurrent with the Corey connundrum, I was aware of surveillance on me. I had strange things going on with my phone, mail, car, and I also had been followed by a black van with dark windows and many antennas on the roof. I had witnesses to these strange occurrences many times. I had never cultivated thoughts of being surveilled or of government conspiracies, but I couldn't help notice there appeared to be an interest in me. I wondered whether this had anything to do with my correspondence with Panda and/or Corey. I decided Andy, my private detective friend, was the appropriate person to help clarify matters. We formulated a plan for him to visit Corey in prison and to find out from the local law enforcement the truth about him.

The truth was not as easy to mirror as we thought it would be. Who was Corey? He was not any of the things he told me. He was nine years younger than he said, he had never been in the armed forces and had been the instigator of a murder. He didn't perform it himself, but got someone else to do it while he watched. He was a very bad dude from the inner city. Ironically, his mother was a clerk in the sheriff's department

What was the point of all this? Andy tried to find out by pushing the envelope a bit. By greasing the right law enforcement palms, he was told in a covert manner that Corey had been involved in a confrontation with the FBI and the CIA. Corey had attempted to scam them by giving them false information related to a federal matter. It seemed that both agencies took Corey seriously. When Corey's information did not pan out and proved to be an outright lie, both of the agencies let him off the hook with impunity.

The local constabulary did not understand that at all. They thought that Corey should have been charged federally, but nothing at all was done. It was being whitewashed, although many people knew about it. Those involved in the legal system who knew Corey was a bad seed hoped he would finally get his comeuppance, but it didn't happen.

Andy thought it was possible that I was being put under surveillance because of Corey's federal involvement. He went to the local FBI office to check with the Special Agents there. He would discuss the matter with them. Andy said that after his explanation about the situation to the Special Agent, he was told to leave the office, they didn't want to talk to him. He insisted at least they check my name to see if I was in their computer. Reluctantly, the Special Agent checked the computer and said that my name wasn't in it *but* she refused to let Andy view the screen! No amount of convincing could get her to do so. With only a modicum of success, Andy left the FBI office.

I told Andy that if he found out anything really scurrilous about Corey that I didn't know, he should gently tell Corey that it would be best if we didn't communicate anymore. Corey then realizing that his game, whatever it was, was over would have to let go. We had found out that he was very dangerous and such a consummate liar that he had even scammed the FBI and the CIA! I wanted no part of him. Andy met with Corey and delivered my message without hostility. Corey knew that we knew the truth of his criminality. We still didn't know why he was dealing with me.

Andy also spoke with the warden at the prison. In checking the tap on my phone line, we knew that it came from the area the prison was in. The warden did not deny that they had been tapping my line. He said it was a perfunctory process and now that things were cleared up with Corey, they would cease the tap. I hoped that candor about the connection with Corey would end the spying on me, but it didn't.

Six months later, I was still aware of unusual effects on my phone. Andy phoned the warden again, who was very nonchalant. He said that he was sure that the tap was off from their side and if there was a tap, it wasn't coming from them.

I hadn't heard from Corey for over a year. Then I received a letter written to me at my home address. Corey asked for a photo he had sent to me of "him and his men who were with him at Mountain Home Air Force Base." I was appalled. I had never given Corey my home address and my address is not listed in the phone book. Corey had always written to me at my P.O. box. I checked with the post office to see if he could get my address. They said that a new law prohibited giving out the home addresses of those with a P.O. box. How he got it, I didn't know. It was now immaterial. I didn't want him knowing where I lived and dropping in for a little visit! He might still have those plans to break out of jail!

I did not acknowledge his letter. I did not want to give Corey any entree to future communication. I also didn't want the prison to think I still was linked to him in any way. Prisons monitor and note all correspondence of their inmates. I was frightened to think that Corey still had me on his mind and had my home address at his disposal. I decided to write to the warden and let him know that Corey got my unlisted address. I wanted it on record, just in case.

The warden's response surprised me. He sent me back my own letter and the envelope I sent it in. His terse and curt response was for me to just not respond to Corey. He jotted it down on the top of my letter to him. There was not an iota of common politeness in his words. Since Corey was a very dangerous criminal, I wanted a hard copy of my complaint on file. The warden sent me back my own letter, so obviously he didn't consider it worthy of even noting it! Not being satisfied, I again wrote the warden of my further concerns asking would he treat his own daughter's safety so cavalierly? This time I got no response at all. I discussed this situation with my attorney. He said that was not the normal way a warden would respond. I then had my attorney write the warden about this subject and once again the warden didn't feel the subject worthy of response. My attorney tried to phone the prison but the warden refused to take his calls.

I don't know why the warden chose to ignore me and my attorney but I can imagine a few possible reasons for this entire situation. In retrospect, there might be more to this than meets the eye. I do not think that it is irrational or paranoid to examine some facts of this communication between Corey and me that bear some scrutiny.

Corey was very young and educationally unsophisticated to have the knowledge of UFO crash retrievals that he had. Where and for what reason did an inner city, street wise punk get these facts? How did he get the military file forms and know how to fill them out? How did he get the metal bracket for the folder? Who helped him send out letters to me from California, Nevada and Texas? He must have been very well connected with other criminals across the country. Why did he have a social security number with the first five numbers exactly the same as mine when we are 25 years apart in age? Why did the federal agents let him get away with his scam, without any criminal charges? Why did the FBI agent not allow Andy to see the computer screen with my name on it? Why did Corey want to send me a television set? Why did Charles Carter call me to infer something illegal? Why did the warden slough me off? There appeared to be a lot of questions and no real answers given to me.

My gut feeling is that this was not about Corey, this was about me! The synchronicity of Corey's and my social security numbers must have meant something. Perhaps we were both a part of some kind of government project. Maybe that was why Andy wasn't allowed to view the computer screen. There might have been some reference to it. Perhaps the federal agents' knowing Corey's expertise at scamming decided to use him to get to me, and made a deal with him. That is why they did not prosecute him. Maybe the television set Corey wanted to send me was not the average boob tube but was for other purposes like mind control! There was some kind of setup or evaluation of me planned. It had to be since there was too much weirdness involved. If Corey, the criminal, had asked me for money or tried to sweet talk me, my point of view would be different. Things would have been obvious. The Corey connundrum remains that, just an enigma.

These possibilities sounded paranoid, but in evaluation of this situation of high strangeness they became rational. The whole of the Corey connundrum appears sane when you compare it to my experiences to come with an FBI agent.

"COREY'S INTENTION IS TO HARASS AND INTERFERE IN PEOPLE'S LIVES. HIS INTEREST IS IN DISRUPTING YOU AND MAKING YOU UNSURE OF YOURSELF. HIS ULTIMATE AIM IS FOR YOU TO QUESTION EVERYTHING YOU BELIEVE. YOU MUST SECURE YOURSELF IN YOUR EXCYLE IDENTITY AND BE AWARE OF YOUR ACTIONS. MOST IMPORTANTLY, WE CAN ONLY HOPE THAT YOU LISTEN TO YOUR INNER GUIDANCE. WE CAN SEND LIGHT TO YOU BUT BE CAREFUL, AS COREY THINKS THAT HE IS INDESTRUCTIBLE. HE BELIEVES THAT HIS MIND IS IN CONTROL, NOT YOURS. THROUGH THE MAIL, HE HAS GAINED ACCESS TO YOUR RECORDS, TRACES YOUR PHONE CALLS AND CHECKS THE CREDIT BUREAU. HE HAS THE RESOURCES TO FIND OUT ANYTHING ABOUT YOU. HE IS ASSISTED BY M.S., WHO IS VERY DANGEROUS AND FOLLOWS YOUR TRAVELS AND WILL TRY TO GET CLOSE. THEIR INTEREST IN YOU IS BEYOND THEIR ASSIGNMENT. WATCH WHAT YOU SAY TO STRANGERS. IT IS IMPORTANT THAT YOU START WRITING AND CHRONICLE YOUR EXPERIENCES."

– Zarg

Ours was a very tumultuous relationship. Tom had just come from four years in an Arab country. He was working for their electric company, ostensibly as a consultant, but in truth, it was a position related to electronic surveillance and the U.S. government was his boss!

THE THOMAS THOMPSON ENCOUNTER

I got into the elevator on the 12th-floor of my condo. I was all decked out in my tennis outfit, complete with racquet and water bottle. I had on my darkest sunglasses and my sun visor. It was not a busy time of day, so I was surprised when the elevator stopped at the fifth floor.

As the door opened, I juggled my tennis paraphernalia and looked up. What, or I should say, who I saw surprised me. He was a hunk. I had never seen him before. I had lived in the building for five years and was a member of the board of directors for the condo for almost as many, yet I had never noticed this guy before and I would have. My first thought was that he was a typical Ft. Lauderdale "Mr. Cool." He was around six feet tall, medium build, slicked back black hair with gray at the temples and a gray-black moustache. He had on trendy aviator shades, so I couldn't really see his eyes. His after shave or cologne smelled great, kind of musky lemon. He wore tight but neatly pressed jeans and a short sleeve silk print shirt. I muttered some descriptive words to myself about his sharp looks. "Nah, he doesn't live here, he must have a girl friend who does and he has just spent the night," I thought.

"Good morning,"I said. Mr. Cool just grunted nothing discernible back to me. He was obviously not a morning person. Well, you can't win them all. The elevator stopped at the lobby but Mr. Cool didn't get out. He proceeded with me down to the garage. Then a weird thing happened. In my mind's ear, I heard very loudly and agitatedly, "Hurry up and get into your car before he does. Don't diddle around with your seat belt. You must be first out before the electronic garage door goes up. He must be behind you." I thought that I must be going crazy. Mr Cool was walking to the same part of the garage where my car was parked. His space was four down the row from mine. Oh, God, he had a green Corvette! How very typically Ft. Lauderdale of him! Even if he was cute, flashy, conspicuous consumption was not my style!

Despite my protesting conversation with myself, I hurried into my car, threw my tennis gear in the back seat and without fastening my seat belt, I backed out of my space and maintained the front position, while waiting for the garage door to open. Behind me was the green Corvette with Mr. Cool gunning the engine. The sound reverberated deeply off the garage walls. I hate all that noise so early in the morning. I was glad when the door opened and we both drove off. I soon forgot Mr. Cool, as I turned off towards the tennis courts.

The next morning, as I approached my car in the garage, I noticed there was note on my windshield. Even from 40 feet away, I knew who it was from. I pulled the note from under my wipers. Written on paper from a very posh hotel in Hong Kong was a cryptic note, "Call me, Tom." A phone number followed Tom's name. I didn't know any "Tom" but I had a hunch that it was from the Mr. Cool in the elevator. I guess he wasn't as disinterested as he seemed to be yesterday. I decided not to call him until the evening. It would serve as a symbolic gesture to all of the Mr. Cools of the world to have to wait to have their demands met!

When I dialed his number, I played dumb. "Someone left a note on my windshield saying to call this number. Is Tom you?" He replied, "Yeah. I was with you on the elevator yesterday when we went to the garage. You drove out ahead of me. I saw that bumper sticker on the back of your car. I really liked what it said. I thought that any woman that has that on her car has to be someone I want to know! I'd like to find out more about it. When can we get together and talk?"

I was taken aback by his response. The last thing I thought Mr. Cool would notice was my bumper sticker! It read "Mind Power," I belonged to a church that teaches an open, non judgmental metaphysical religion which espouses basically that all is created and connected out of the mind of God. We see God as an infinite, divine and creative Intelligence. What was really interesting was that Tom noticed the bumper sticker that I had created and donated to sell to church members and friends. I then remembered the urging from "somewhere" in my mind's ear, to hurry and get ahead of Tom's car. Now, I understood why I had to do that, so that he could read that bumper sticker!

Was there some kind of pre-ordained destiny involved here? Were the Fates orchestrating our lives? What purpose would this serve? This was getting very interesting. Obviously, it wasn't just your average pickup line at work. How could I say no! He was kind of smooth, anyway, even if he did have that flashy car! So, we planned to meet and see what course was set before us.

Ours was a very tumultuous relationship. Tom had just come from four years in an Arab country. He was working for their electric company, ostensibly as a consultant, but in truth, it was a position related to electronic surveillance and the U.S. government was his boss! Tom swore that he would never do that kind of work again, especially in the restrictive environment of an Arab country with no booze or women easily or legally available! He attributed his prematurely gray hair to the experience. He said that his food had been salt petered to make sure he didn't get into any trouble with the locals!

How a guy like that could be interested in my bumper sticker was truly paradoxical! But he was more than just a pretty face and a limited personality. He was interested in metaphysical and paranormal things. He said that he was born with a "veil" or a membrane over his face. People who have had that are said to be very psychic and have unusual abilities of omniscience. I guess that was why he was interested in my "Mind Power" bumper sticker. He was also an achiever and had surpassed all in his family in education. He was an avid reader and prided himself on his stock market expertise. Even though, we had an undeniable attraction for each other, our personalities were mismatched. After a year, he moved to a nearby town. We spoke often on the phone but it never was the same. He had a difficult time finding a job that yielded him the cash he was used to making with the government. He left the area to sell time shares and eventually ended up in New Orleans. After a few weeks there, he phoned to tell me that he was leaving the country in a few days.

Tom said that a guy walked into his time share office to look at the apartments. During conversation they found out that they both had done the same kind of work for the government, electronic surveillance. The man said that he was working at Kwajalein, Marshall Islands on the Star Wars Project and that they needed more help there. He suggested Tom apply and Tom did. I was very surprised. Tom had been very adamant in his conversation with me that he would not do that again, and here he was volunteering. I also thought that it was kind of strange that a man in that unique job category would just happen to pop into Tom's life like that. Nonetheless, within one week, Tom was employed by the U.S. government on an atoll in the Marshall Islands, just a pin point in the Pacific Ocean.

This was the time in my life when I was a neophyte ufologist. I had just begun to understand my personal connection with the cosmos and was researching the UFO/ET connection with a huge passion! I never mentioned word one to Tom about this subject matter. He was just another part of my world and not one I was certain would understand. Ironically, Tom's letters started to suggest that I read books on UFOs. It was an interest he said that he was cultivating. Since we then shared the interest, I felt it a safer subject and told him I was now a researcher myself.

I went to a metaphysical conference in Barbados. There were many interesting speakers. The one I most looked forward to hearing was J.J. Hurtak, the author of The Keys of Enoch. It is an epic tome about the evolution of the planet as it relates to the spiritual and interdimensional hierarchies. It also maps out the points on the planet of energy grid areas for space-time transcription and protocommunication. The grid area I found most interesting was

Kwajelein, Marshall Islands. It probably was no accident that the government put a Star Wars facility in such a remote out post. It is not a secret known only to our off-planet visitors that an appropriate energy grid is that tiny point in the midst of a great ocean.

I also wondered now about Tom. We certainly met in a most serendipitous fashion and now we shared kindred interests. Was this by accident or by design? Was this some kind of Karmic burn off for both of us or for one of us? After several years of Tom's not delivering on his promises to meet, I decided I no longer wanted to be his pen pal and stopped writing him.

He subsequently went back to doing the electronic surveillance in the same Arab country he worked in before we met. I met Tom before my awareness of the surveillance on me by government agents due to my UFO research. Was this in some way connected? Did the government know about me before I was aware? Was Tom in on a reconnaissance mission related to me? Did he tell them about me? Was it one of these, a combination of these or none of these (Though, *I* was the one who had the intuition to move my car out first.) So maybe it all was a coincidence! If so, it was one hell of a coincidence!

The level of high strangeness in the everyday life of Professor Edith Simpson seems to be more than the average person can tolerate or endure.

THE PROFESSOR
AND THE PROFESSOR

Edith Simpson is one of those special teachers that we remember. She knows how to stimulate an often difficult subject matter and make it palatable and digestible. That is why I heard about her. Simpson is a professor at a southern college in the department of Earth Sciences. My friend, Wanda's, son was one of her students, probably one of her more favored students, since Wanda also had a social relationship with Simpson. When Wanda heard about my interest in UFO research, she suggested that I meet Edith Simpson. She said that Simpson, in an attempt to keep students' minds open to the infinite possibilities in the cosmos, told an incredible account of her personal experience with the unknown. Simpson said that in 1947, she had seen with her own eyes, the bodies of non-human occupants of a spacecraft and the remnants of the craft in which they had crashed. Of course, I asked Wanda to set up a meeting with Edith and to do that as soon as possible. It took over three months until we met.

We met at a restaurant that Professor Simpson favored. It was across the street from her home. She had been going there for many years and knew everyone. Simpson brought along her long-time friend, Marge. My friends Susan, Wanda and I completed the dining group.

Professor Simpson is a tall, sturdy woman, 64 years old. She has a commanding, pedagogical presence. She is definitely not a wimp. She uses language very precisely. Accustomed to the paranoia of UFOlogy and witnesses' fears of being seen or heard in public, I was amazed at Simpson's lack of paranoia. I had my tape recorder and microphone out and asked if there was a more private space to hold our conversation but she insisted that it was perfectly fine to interview her right in the center of the dining room, while we were being served. She made no attempt to lower her voice or mince her words. It was her ballgame and ballpark. Waiters and other diners must have overheard our conversations. Apparently, Simpson did not seem to care and she had quite a bit to say.

Edith Simpson was a very bright student, probably brilliant. Her interest was in the sciences. In the summer of 1947, she was one of the university students chosen to study advanced physics with a world famous scientist at an eastern university. It was a dream come true for her. The time spent with him that summer was more of an experience than she had expected. "The Professor," as she called him, was very warm and friendly to all he met. He especially took a liking to her.

While she was with The Professor that summer, he was summoned by the U.S. Government to come to an emergency meeting, a gathering of the elite in many areas of sciences and military, to a place in the southwestern or western United States. Edith Simpson, 18 year old student, was asked by The Professor to go with him to his special meeting. Since all of the students that were studying with him that summer needed a security clearance, she already had one and was allowed to attend the meeting with him, as part of his entourage. She had no idea, at the beginning of the trip, what would ensue.

How Professor Simpson related the details to me, I believe, was more than interesting. She did not expect such an intense inquisition on my part. At first, her responses were somewhat limited, at times cryptic. Often, she would return to a past vague statement and embellish the details, as if some kind of veil was being lifted away. Sometimes she would say, "I haven't really thought about this in a long time," or "I don't understand why I don't remember this now, I should know this!" The more I prodded her and returned to past questions, the more she seemed to have definitive responses. Her eyes reflected brightly with every newly remembered data byte. I wondered whether she was really remembering or just giving me what I wanted to hear. Probably, the answer to my own internal query came in further questioning.

I asked Simpson if after she returned to academic life she ever had any follow-up phone calls or contacts with any government agencies. At first, she replied that she didn't think that she had although soon after she returned she had a feeling that someone was listening in on her phone calls, but it was just a vague feeling and nothing came of it.

Then, Simpson recalled an experience she said she had forgotten but upon my questioning just remembered. She said that she was called out of her college class by the Dean. He introduced her to a woman that he called a "psychologist." He said that the "psychologist" wanted to ask her some questions and that she should go with her. Edith followed her to a cleared out utility closet. There were just two chairs in it. They sat down and the "psychologist" proceeded to ask her one question, "As a child, did you walk or talk first?" The next thing Edith could remember was that she looked at her watch and it was an hour and a half later! She couldn't recall anything that had transpired in that hour and a half. Professor Edith Simpson said, "It was just as if she had hypnotized me, although I had nothing to hide but I just didn't know what the probe went to."

I thought that off-handed statement about the "psychologist" could be important. It might prove the beginning of a mind control scenario, one way or other. I asked whether she had ever had any buzzing or ringing in her ears currently or in the near past had any strange phone calls at regular intervals. Simpson looked across the table at her friend Marge and they shared a knowing glance. "Oh, yes, I often have had phone calls where no one was responding." At first, Simpson said that she thought that they were just "wackos" who were interested possibly in breaking into her home. But then she continued, "I thought for several years that maybe the government wanted to know whether I was telling anybody about what happened and they'd listen in and I sensed that. But that was just my own perception and has no basis in fact, whatsoever."

Professor Simpson had a penchant for ambivalence on the matter of government concern about what she knew. At first, she denied that there was any interest in her and then she indicated the polar opposite of that position. On one hand, she felt she was nobody important and in the next statement she stated an experience of feeling surveilled. She stated another incident while she was working at her current college where "...a couple of Federal men came by one day and asked a lot of general questions about me to different people. I wondered what they were asking those kinds of questions for."

Professor Simpson revealed that she had been having continual problems with the IRS since 1975 that had not been resolved yet, even though she had affidavits from many witnesses regarding the existing problems. Then there was the unending scenario of disputes she has had with her local municipal government related to her home property. She was in court over these at the time of the interview. In the last few years, her home phone service had been regularly interrupted. For three months prior to our meeting, there had been a constant busy signal on her line. She had not been able to use her phone. The phone company said it was a problem within the home lines and Simpson maintained that it was the Bell System's. No matter, she was incommunicado, phone-wise, and was only reachable at her school's office.

Most bizarre was Professor Simpson's being the one woman victim of a major crime wave. In the previous seven years, she stated that she had been subjected to six home invasion burglaries, 15 muggings, (we were shown scars to corroborate this fact) and numerous car thefts and break-ins. Professor Simpson's friend Marge shook her head in agreement to all of this terror. I asked Marge, who is older and more fragile looking than Simpson, if she had encountered the same problems. She replied that she just had one mugging only! Showing my incredulous reaction at the large number of criminal events in Edith Simpson's last few years, I said, "Don't you think that is an outrageous amount of crime directed at you?" Professor Simpson shrugged off her response with, "This neighborhood isn't what it used to be." My friends and I looked at each other and knew that there must be more to this than just a neighborhood in transition.

There certainly was. Perhaps the reason for all of this crime was the fact that Professor Simpson said she had in her possession 48 35mm photos of what she saw with The Professor in 1947. She said that the "scientific group" that was there had access to them. The spaceship and the dead aliens, in both full figures and parts of bodies, appeared in them. I asked her if I could see the photos and Professor Simpson replied that most of them were no longer in her possession. She said that her car was stolen from a large shopping mall and that the briefcase in which she carried most of the photos was in the stolen car. Simpson said that the car and the photos have been found and retrieved but that the police refused to release them. She said that the police told her that they had the photos even though she did not list the contents of her briefcase on the police report. She was afraid to say what the photos were, due to her position at the college and in the community. Although when the police told her that her car had been found, they also mentioned they had found the photos and made sarcastic remarks about the subject matter. She said that she was afraid to make an issue about the photos, since she was having so many problems with the local municipal government but would take up the issue of their return, once her problems have been resolved.

The level of high strangeness in the every day life of Professor Edith Simpson seems to be more than the average person can tolerate or endure. There was a period of missing time in Simpson's life that occurred about five years ago. One Friday night, she went to sleep and was awakened the following Monday morning by her friend Marge and Marge's husband, Allan. The couple had tried to phone Simpson the entire weekend but with no response. They feared that she was dead, since her car was in the driveway and she was expected to be at home. Allan had to break a window and enter her home that way. They found her asleep in her bed. Professor Simpson said that she could not remember anything about that weekend, including getting up to go to the bathroom. She said it was a total blank in her mind. She supposed that she might have had the "flu or something." In my attempt to corroborate Professor Simpson's statements, I had a number of dead ends. She was vague on specific dates and having a very common first and last name, the computer lists of incidents of crime are difficult to penetrate. I know for sure that she is, indeed, a professor of long time standing at the college. She is also well-known and has resided in the same community for 50 years. She showed no anxiety about discussing the issues of 1947 and The Professor she adored. If she did experience the viewing of the aliens and the craft with The Professor, it is very possible that she might be under mind control and surveillance by governmental agencies and has been since the viewing first occurred. This would explain the session with the "psychologist", the "lost weekend" and the vagueness about dates things occurred. It also may relate to the municipal, IRS and telephone problems. But if she truly was under mind control and pressure, why would she even mention her experience, let alone say that she had photographic evidence to substantiate this?

If Professor Edith Simpson was not telling the truth, what would've been her motivation? She had a lot more to lose than to gain. She was an esteemed member of the community and had a job where being a weirdo was not well tolerated. She carried herself with dignity and was very articulate. If she was nothing but a nut, she could easily fool a lot of people into believing that she was telling the truth. But then there was also the chance that the "powers that be" wanted us to *think* that she wasn't playing with a full deck and *wanted* us to feel that she lacked credulity! Simpson knew some details about a 1947 crash case that had not been published, although there were still many missing pieces in the puzzle that she created for us.

WORDS ON A WINDOW FRAME:
A WHIMSY OR A WARNING?

On November 23, 1993, 1 had my last phone conversation with Professor Simpson. It was not satisfactory to me. She had a litany of problems that led to excuses as to why she could not provide "copies" of the photos we requested. The problems she enumerated were so many and so burdensome that it seemed more than one sixtyish woman alone could bear. At the end of our lengthy conversations, she did what she always had done, she promised our satisfaction but sometime way down the line. After all of the months of dealing with her, I did not expect anything more of her.

Little did I anticipate that I had a surprise on the way. On November 24, 1993, after I had completed an early morning tennis match, I went to answer the phone in my den. As I was seated at my desk, I turned to the windows that faced directly out at the ocean. The windows are modern awning-type that have metal frames. They encompass three windows together, each with four panes that open in an awning-type fashion. In the center window on the center metal frame, I thought I saw some kind of writing. I went to the window and couldn't believe my eyes. There deeply etched into the metal were the words *come home my love* and what looks like the initials *"O.H."*. I was shocked. I tried to rub the words out but they wouldn't erase, they were engraved into the metal. I look out of the window every day that I am home. It is my favorite window to stand in front of, since it has the best view. I had lived in the apartment for 13 years and never saw this writing before and it would be hard to miss! I check out the weather there everyday, especially the days that I am playing tennis. Why and who would do this? How and when did they get into my home?

My first thought was to check with the cleaning person, Maria, who worked for my condo and also cleaned my apartment and windows every two weeks. She had been doing that for eight years and was absolutely trustworthy. I wanted to know if she had ever seen this before. She had just washed the windows five days prior. She said that she had never seen the writing before. I then beeped my friend, Don, who had remodeled my apartment nine months before and knew every inch of the place. I had him inspect the engraving and he, too, said that he had never seen that before and would have noticed!

The fact that there was no evidence of breaking and entering really bothered me. There also was nothing missing or out of place. That meant to me that the intention was not criminal even though the act was! How did the person or persons get into my apartment? The only people who had keys besides myself were my mother and Maria. I live in a building that has 24-hour security guards at the main entrance but for the days from November 22-24, the service entrance was fully open and unlocked, due to the laying of hallway carpeting in the building. It would be quite easy for someone(s) to slip into the building and ride the elevator to my 12th floor apartment completely unnoticed.

What I found most ominous was the message engraved into the window frame, *"come home my love"*. It certainly sounded chummy, kind of like a unrequited lover or a languishing one. To my knowledge, there was no one in my life that fit that stereotype. Romance had not been my focus, since I got involved in UFOlogy.

The logical progression then was to assume that this was linked to the case of Professor Simpson. I am sure that the verbiage used in the message was meant to be construed by the non-cognoscenti as innocuous, nebulous and shoulder shrugging. If it was threatening, it would be more damning to the agent(s) involved and lend credence to the illegal entering of my home. I decided to make a police report on this, even though I knew that it would be considered a non-event to them, as there was no real evidence of a crime, other than my say-so. The police officer took the information and it was stated on the police report as a "Miscellaneous Incident." The officer said that in over 22 years of his time on the force, in the area, it was the first time he had ever been to my building for any kind of an incident. I agreed with him that in the 13 years that I lived in the building, there had never been a burglary by an unknown individual.

An important postscript to this related to possible phone surveillance on others who were connected to me. After this event, I decided to phone my good friend, confidante and private investigator, Andy. When I told Andy what had transpired, he thought it quite bizarre, even for agents. He also mentioned that he had been having anomalous drops on his phone lines. He would lose phone calls in mid-conversation. His contact within the phone company could not understand where this draw on the line was coming from. After two weeks had passed, Andy stated that he and his wife had tried to phone me from their home and could not get my phone. He said that a recorded phone message would come on saying that my phone did not accept incoming messages, or there was a continual busy signal, despite the fact that I had call-waiting. No other people had indicated the same problem with my phone.

Concurrently with Andy's phone problems, my friend, Don, who lived in another town and who I had beeped initially about the engraving, also had immediate problems with his phone. He could not get any calls out and only one call in, on his phone. He was very upset since it included Thanksgiving Day, when he wanted to phone his family. When he checked with the phone company, no reason for the problem was given. The service interruption disappeared after two days.

My conclusion to this event of very high strangeness was that there was no doubt someone(s) unknown illegally entered my home and defaced my property with the intent to intimidate me. The synchronicity time-wise, it appeared, related to the completed case of Professor Simpson. Why "they" waited for completion is questionable. Perhaps, it was meant to be a permanent reminder of the eternal vigilance of the agents.

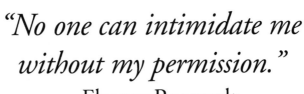

"No one can intimidate me without my permission."
– Eleanor Roosevelt

OBJECT OF SURVEILLANCE

It has become quite commonplace and even trendy in today's world to cast aspersions on the United States Government. Since I have always been of a more conservative or centrist viewpoint politically, government condemnation was not my rallying cry. Despite the growing problems we face in this country, I do think we have the best system on the planet. I am very proud and grateful to be an American. I vote in every election and make contributions to the best candidates in both parties. But I do take umbrage with some individuals who are not elected by anyone, but are employed within our system to serve the citizens of our country.

My complaint therefore is *not* with the government but with those who hide within the bowels of the bureaucracy and violate the constitutional safeguards of our legal system. They are for the most part, nameless, faceless individuals who run rampant with their power and violate our rights of privacy with impunity.

I am a patriotic citizen. I pay my taxes and don't commit any crimes that I know of. It has been the shock of my life to have come to the conclusion that I have become an object of surveillance by certain intelligence agencies. I am still trying to figure out why they have chosen me for their scrutiny. When I first came into UFO research, I heard all of the tales of harassment, surveillance and even worse, done to investigators, abductees and witnesses. Categorically, in the beginning, I did not believe any of it. It was against everything I knew to be true about my country and was contrary to my belief in the system.

Therefore what has transpired in the last seven years of my life is still difficult for me to fathom. As evidenced by the words on these pages, since my interest in UFOs evolved, my Constitutional rights have been violated many times. I do not say that easily or without careful consideration of the evidence I have compiled. The surveillance has been ongoing and varied. I have witnesses who can substantiate my claims and an audio cassette library filled with phone call oddities of very high strangeness. Witnesses have seen me followed by a dark van with blacked out windows and antennas on the roof, heard strange phone calls with an assortment of beeps, tones and cryptic messages, seen mail from my attorney opened and resealed and forwarded from another state. Even my very straight and upstanding accountant stated I had an unorthodox audit of my tax return. He called it "highly unusual!" And most recently, I had a direct contact with an FBI Special Agent who had previously worked for Naval Intelligence and the CIA.

I did everything I could to make sure that I wasn't just imagining things and becoming part of the paranoia of ufology. I hired a private detective who unfortunately validated my conclusions and then became an object of surveillance himself. The worst part of this situation, for me or any citizen, is that no matter how much evidence you amass, all of the evidence is anecdotal! My private detective had an inside contact at the phone company who verified and saw the phone taps for himself. Phone taps require paperwork with official judicial orders and leave a paper trail. We were able to follow up on one but two others led nowhere. There were no paper trails or judicial orders. Just the taps, going who knows where! Currently, the state of the art in surveillance, including phone surveillance, doesn't even need the phone system to listen in on. There are other high tech ways of doing this. We all are vulnerable.

I have had to devise my own primitive method of surveilling the surveillers. With the use of answering machine, tape recording my phone calls and the caller ID service, I have at least been able to chronicle and log in strange phone calls. I have a lot of evidence but nowhere to present it, without looking like a nut! That is a big problem for us all.

At this point in time, I cannot say that I have ever been threatened but there has been a concerted effort to intimidate me. It is interesting that many times there is no effort to be covert. "They" want me to know that they are watching and listening. It is a stupid and childish performance. It would be far simpler to just ask me the questions they are seeking answers to! It certainly would be a lot more cost efficient, fiscally. As a good citizen, I resent this waste of my tax dollars!

If all of the listening and watching relates to my interest in UFO research, this only serves to validate the field of inquiry for me. Something must be going on, otherwise they would just ignore me! That is the only positive value in this for me. I know I am not wasting my time!

Above all, the pervasive unmerited and unrestricted violations of the constitutional rights of privacy on innocent citizens is the first step in the breakdown of democracy. If it could happen to me, the "Watchers" could be looking at you, too! And I don't wish that on any good citizen.

It seemed that the island of Puerto Rico was Earth's 20th century space port and one of the best kept government secrets, and the US Navy was an integral part of keeping that secret.

SPACE BASE PUERTO RICO
"WHERE THE G-BOYS ARE"

Science fiction has become science fact in Puerto Rico. Whatever UFO phenomena the mind can conceive has been, and still is, taking place on that small island paradise surrounded by the Atlantic Ocean and the Caribbean Sea. Extraterrestrials, interdimensionals, spaceships that appear and disappear, abductions, encounters of the U.S. military with "alien craft," bases inside mountains, under the earth and under the ocean and lakes, hybrid children and adults, apparitions of the Virgin Mary and sun pulsations have all manifested in generations past and at this moment in Puerto Rico. Quite often, families have had these experiences generationally, as if they are being tracked like tagged animals. It has taken its toll on the population but it seems that these phenomenon have inured the locals and are now met with just a shoulder shrug of acceptance! Puerto Ricans will often admit, once they trust you, that they have had or seen most of the aforementioned scenarios or know someone who had. The basic nature of these islanders is kind and friendly. They are an amalgam of hispanic past and American present all wrapped in a very interesting package. I love the island and feel quite at home there. I feel grateful for the opportunity to have researched the unique nature of the experiences only Puerto Rico can offer in such abundance.

In 1991, a dozen of our local South Florida UFO research group travelled to Puerto Rico for a first hand look at what was going on. Fortunately, we had the company of excellent local researchers who shared their knowledge of past and current UFO and US military happenings and introduced us to abductees and witnesses. Language and culture were not barriers, since half of our troop were of Puerto Rican lineage. The trip was never boring. We had a lot to do, see and experience for ourselves.

Our first stop was the popular seaside town of La Parguera. Since most of us were on a budget, we stayed at what probably was the cheapest and crummiest motel in town. It was in a great location on the main drag but the amenities were definitely spartan. The nicest thing about this place was the outdoor vine-covered patio area with lots of comfortable chairs. It was right in front of the check-in area and next to the hotel parking. We gathered there to wait for Juan Marron, the best known researcher on the island. He was coming from San Juan to introduce us to local abductees and witnesses. While we waited, we were having a great time chatting, listening to the music on the audio system and enjoying the sun.

Suddenly Elena, a Puerto Rican abductee and a member of our research group, told us to look at the car that had just parked nearby and the two men who were in it were now moving directly towards the motel check in area and us. She said that they were government agents. Both of the men were in their late 20's or early 30's. They were very clean cut, preppy looking guys. One looked Latino and the other Anglo. I asked her how she knew that they were agents. She said that on the back window of the car, there was a gray, waxy looking seal about three by four inches in size. The seal has an eagle impression on it. Elena didn't know which agency it was, but she had learned that agents' cars sported this decal so that the local police would not bother them and would let them go about their business. Elena said that she had seen this seal before on government agents' cars who had investigated her experiences and tried to harass her.

Elena was a very special person. She told us many incredible stories of her childhood when her mother, who also had extraterrestrial involvement, fed her mercury on a regular basis. She also told of a space craft that landed on the lawn at her home in Yonkers, New York. I found these anecdotes difficult to believe, as they were so outrageous. But, Elena was forever credible to me after her observation about the two young men we saw that day. This was a woman who knew what she was talking about. She proved to me her experienced insights were unimpeachable!

We observed the Anglo man going to the check in desk. Estella, another Puerto Rican with us, was standing at the counter looking at post cards in a rotating rack. She was within inches of the hotel employee at the desk and the Anglo agent. Estella saw the agent show the desk clerk some identification and heard the agent tell the desk clerk, in Spanish, that he and his co-worker were at the hotel to protect a very important political figure who was staying there. The agent then asked the clerk to provide the names of all of the people who were registered at the hotel. When Estella heard that she ran over to where we were sitting and we could see that she had something to say to us covertly. She was quite agitated. At the same time, the Latino agent stood right near where we were all sitting.

We were discussing the name of a Spanish song that was playing on the sound system. The agent corrected what we thought was the name. We were astonished that he should be that bold. When both of the agents walked away from the hotel, we finally were able to discuss what had occurred. Estella told us what she had just heard at the check-in desk and we told her that we already knew they were agents, since Elena had pointed them out and showed us the seal on the car. Now the big question was were they there to check up on us? We could not imagine that any very important politician would ever stay at a motel where you couldn't flush the toilet and turn on the faucet at the same time! This place should have been called, "El Dumpo!" We did not have time to ponder the question any further, as Juan Marron and his friends arrived to pick us up. We got into the van and drove off to visit an abductee.

Mauro, the abductee, owned a shop catering to tourists. It was very tiny and had only limited air conditioning. We decided that half the group would hear his story while the other half waited outside. While we were waiting, we got some cold drinks and stood on the sidewalk talking about the day's agenda. As we were chatting, a car drove by slowly, with its two occupants looking directly at us as it cruised by. We recognized them as the Anglo and Latino men we saw at the motel. They made no attempt to hide their obvious interest in us. We decided to make a joke of it and waved at them. When we lost sight of them, another car stopped in front of us. It was the same

47

make and year as the young agents' car. It even had the same car dealer's tag from Miami that the agent's car had. This time a middle aged woman in glasses was driving. She turned towards us and asked, pronouncing each word distinctly, "Did you see where the agents went!" This was too unbelievable. What childish and stupid antics on the part of the government team! We looked at each other and laughed. In unison, we said, "Yeah, they went thataway!" and pointed in the direction the agents were heading. We were incredulous. If this was supposed to be "undercover," they had better go back to spy school! But, then again, it appeared they wanted us to know that they were around and aware of what we were up to. As far as I knew, we had not broken any laws, nor did we want to. To make matters even more ridiculous, the two male agents again drove by us slowly and turned to once again check us out. And once again, the female, probably another agent, drove up and asked the same question as to the agents' whereabouts.

We all decided what a waste of time and money this type of surveillance was and if it was meant to intimidate, it did just the opposite. We were well within our Constitutional rights of inquiry and free speech.

We never noticed those agents again. However, on a beachside night vigil at a well known UFO viewing spot near Cabo Rojo, we were told by a friendly local police officer that we had to be very careful and not walk down the dark road alone. He said there were people who knew what we were investigating. He went on to say that they did not like the fact that we were there, so we should be careful. The officer then went on to tell us of his own numerous UFO sightings during his rounds on this isolated road.

We all knew that the officer was not warning us about average Puerto Rican citizens. Most of these citizens were happy that finally all of the activity in the island was getting some attention from statesiders. They wanted the world to know what was going on in their little paradise.

Our group was also fascinated with an experience had by a police officer and his family during a visit to El Junque, the rainforest area in the northeastern part of the island. It is an incredibly verdant paradise which becomes more lush the higher up the mountains you go. El Junque has always held a special significance in the affections of the islanders as a magical place. The magic has given way to terror recently with hair-raising tales of resident aliens, UFOs and secret US government installations deep within the heart of the rainforest.

The police officer and his family took the proverbial wrong road in an attempt to get to the other side of the mountain. As they became aware that they were lost, they saw what looked like a guardhouse and thought they could get some directions there. Moving the car slowly towards the guardhouse, they were shocked at what they saw. Standing sentry-like in front of the guardhouse were two Robocop-looking beings. They had dark helmets with reflective visors and totally black jumpsuits. In their gloved hands, each "guard" had a wand held as if it was a weapon. It was a totally awesome and fearsome sight. Before they could utter a word, from behind the guardhouse came a truck that had "US Navy" on it. Two men in fatigue-like jumpsuits got out of the truck and ran over to the lost family's car, yelling at them as they were approaching. They said, "Get the hell out of here! If you stay here one more second you'll be in big trouble. Turn around and get out now!"

In maximum anger and fear, the policeman turned his car around and found his way back down to the road, too terrified even to look back to see if he was being followed. When they finally were able to stop the car and evaluate what they had been through, they all felt that something, possibly even death, would have happened to them at the hands of the "things" in the helmets. By the way the Navy guys were acting, it seemed apparent that something not normal was going on, on the mountain.

Our research group decided to take our own tour of El Junque. We wandered up the main roads filled with tourists and then decided to look for less travelled side roads, in hopes of seeing something as paranormal as the many experiences we had heard about. We found a road that had a locked gate at the end of it and a sign that said "US GOVERNMENT - KEEP OUT - RESTRICTED AREA." As we were standing there trying to decide if we were going to be stupid and sneak in, a Ford Mustang approached. It had Massachusetts license plates on it and a woman at the steering wheel. "Can I help you?" she said.

Playing dumb, I replied, "We thought we had hiked this road before and were wondering why it is closed now?"

"No, this is a federal installation here, you'd better hike elsewhere." Brazenly, I continued to probe. "Oh, do you work here, I see that you are from Massachusetts?"

"Yes, I do, I am a biologist for the US Navy."

Another member of the group, equally brazen said, "I have heard some strange things are going on around here, have you seen anything?"

Without hesitation, she replied, "I have worked here for three years and I have never seen anything. It is just a bunch of local folklore. Well, I better go and you guys better stay on the normal hiking paths." She then got out of her car, opened up the gate and said goodbye.

We all thought that was a very worthwhile fishing expedition, and we didn't even bait the hook! We confirmed for ourselves that the US Navy does have sites deep within the rainforests. What could their purposes be? Why would the Navy find its way so far from the water? There were other naval facilities like Roosevelt Roads and Vieques on the islands. What brought them here? We didn't even have to mention anything about little gray men or flying saucers, the biologist knew just what we were talking about. "Local folklore," she said.

"Bah, humbug!" we thought.

Why a Navy biologist in the rainforest? It seemed that the island of Puerto Rico was Earth's 20th century space port and one of the best kept government secrets, and the US Navy was an integral part of keeping that secret. With all of the local action going on, it looked like even Robocops and their magic wands wouldn't be able to keep the truth from concerned citizens. Arnold Schwarzenegger, be warned, you've got some heavy duty competition out there!

How the real life 'X-File Man'
found me, I don't know.
Knowing the level of coincidence
in my life, the answer could
only be moot.

THE CONDO-MANIAC AND ME

Living in a high rise South Florida beach area condominium is a mixed blessing. It is wonderful to be near the energy and light of the ocean. The incredible, ever-changing panorama of water and sky supports my aliveness. What does *not* add to this enriching life style is the infernal cacophony of a few neighbors. Luckily, they are very few but the bad neighbors are very bad!

In the years since I have lived here, I have tried hard not to be on the board of directors of the condo association. No matter what good works the volunteer board does, it never satisfies the constantly crabby few. They are chronic complainers whose lives center around being unhappy. Reluctantly, I have served several terms on the board to preserve and protect the sanctum sanctorum of my home. I didn't want certain inmates to take over the institution. Throughout my tenure on the board, there have been few monumental battles and a lot of silly skirmishes. But all this experience aided in building my mental and emotional armor, readying myself for the most valiant of campaigns, the one against the condo-maniac of all times, Mark, the former FBI agent!

The other board members and I felt that we must protect our home from Mark's unabated verbal assaults and threats. His continual barrage of illogical and unfounded complaints to the Bureau of Condominiums was creating a most unpleasant and fiscally damaging situation for the owners and residents. He never sought to ameliorate differences, only to escalate them. There was no way to appease the man and his tiny band of malcontents.

It seemed that Mark's disruptive behavior began after he "retired" from the FBI. His whole world and identity appeared to be built around his job. He let everyone know where he worked at every opportunity, even after he retired. While he was employed with the Bureau, he even walked around with his weapon visible in his shoulder holster. I don't know whether that was meant to impress or intimidate. I do know that is not normal for most agents to do this. For their own personal and family's safety, they rarely seek to draw attention to their profession.

The entire process of Mark's terror tactics strengthened the bonds of the board. We felt we had a fiduciary to perform to protect the owners' interests. We also were concerned that the board might become a target for his venomous revenge. We decided to do something about him. Mark, himself, provided the appropriate vehicle. During one of his tirades, he wrote a letter to me. It was really bizarre. He said that he and his wife were working in cooperation with a Congressional Investigation committee in a surveillance operation on our condo premises. He continued by saying that he and his wife and other armed personnel were conducting sweeps on a regular basis.

When I read his letter, I knew that Mark had definitely crossed the line. How could a sane person who so proudly proclaims his former FBI service to everyone, write such a profoundly outrageous letter that was an obviously patent lie? We checked to see if he had a gun permit and he did not. So there was no way that he could be legally involved in carrying a weapon. We were all becoming concerned about his mental state. Something had to be done.

It was the consensus that we must let the FBI know what their retired brother was doing. The "image" of the agency is always important. A loose cannon like Mark could eventually make them look bad. Whatever the reason he "retired" from his job so early in his career, he seemingly found no comfort in it. And neither did we!

With the consensus of the board, I contacted my private detective friend Andy. He had many law enforcement friends at all levels. Andy took a copy of Mark's letter about being a member of a Congressional investigation team, and began to make some inquiries for us to find the right people inside the FBI.

Eventually, our efforts paid off. Through Andy's contacts, the right man appeared. Jordan, the FBI Special Agent, had worked in the same office with Mark. It appeared that he knew him well. After a few weeks, Jordan phoned me. He requested that we meet to discuss the problem. He preferred to meet somewhere outside of the condo, fearing that he might meet up with Mark. We decided to meet at a local restaurant the next day. Jordan told me what he looked like. He said that he would be carrying a fanny-pack, as that was where he kept his weapon. He sounded very polite and professional. Considering my past experience with personal surveillance, I had mixed feelings about this. I asked if I could bring one of my cohorts, but he said that at this time, he preferred it to be just me.

The next day, I met with Jordan. He was just as he described himself. Impeccably well groomed and soft spoken, he seemed more like a professor or an accountant. We took a booth in the far end of the restaurant. He showed me his credentials and we began the discussion. Jordan was an excellent investigator. He asked all of the right questions. I was impressed with his technique and I told him so. He responded with modest "thanks".

At the end of an hour's discussion about Mark, he asked me what "Jackson" was. I looked at him with amusement. "That's my maiden name, I guess you have been checking up on me before you came here. I did say you were a good investigator."

He then replied that he wouldn't ask me my age, because he already knew that. I smiled at him, "What else is in my file?"

"You don't have a file," he replied. He looked up at me and without missing a beat said, "I like your wrist watch."

"Oh, no," I thought, "what am I in for now?" I was wearing a watch that had been on my wrist almost every day for nearly four years. I had bought it at a UFO conference in Gulf Breeze, Florida. The watch face had a drawing of the "Gulf Breeze UFO." It is a lantern-shaped craft that had been seen and photographed in that area. Very few people, who were not directly looking at the watch, had ever noticed the specifics of the design. I was not sitting that close to Jordan for him to notice. If he *did* notice this on his own, he truly *was* a great investigator.

I was stunned at his comment. I was beginning to wonder what the true nature of this interrogatory would be. Was this guy just killing two birds with one stone? I responded that I was interested in the UFO subject. Quite incredibly, Jordan said that in his FBI office, he was known as the "X-Files Man!" It was his job to investigate MIB (Men in Black) complaints and occasional alien abduction reports. Quickly, he added that his FBI turf related to investigations of impostors, so this was part of his domain.

I was a bit stymied as to how to pursue this conversation. His demeanor seemed to change, when we proceeded with this discussion. He was more relaxed and chatty. Was this a guise or was he really interested? He stated his skepticism about the whole gamut of ufology, with the same rationale and facts that the normal debunker would throw into the discussion. I didn't force the issues but decided to discuss them in a limited way, to see if it would lead me somewhere.

It was now more than two hours since we had walked into the restaurant. I was amazed that he was not pressing to end the discussion, since law enforcement types are generally not interested in prolonging time spent, unless they want to get somewhere factually. I told him that I wanted to show him something on my car that I thought looked like vandalism. Jordan's car was parked next to mine. I showed him the damage on my car and he again returned to ufological talk. We had already shook hands to say goodbye but we were not leaving. We each stood next to our open driver's side doors and the discourse continued.

The conversation moved to the relevance of the Marianist apparitions and the effect of mind on the Super Collider! I think he was a little amazed at me. I was amazed at me! I almost forgot that he was an FBI Agent, but he seemed sincerely interested. After another hour of discussion in the Florida sun, with the sweat rolling down our necks, Jordan again came over to shake my hand. He said that we should continue the discussion at lunch another time, since he was sincerely interested, although skeptical, about the subject matter.

I smiled and agreed that it would be nice. He said that he wanted to discuss Mark some more.

The next day, Jordan phoned. He said that he had found a safe way to get into the condo without being noticed and if it was convenient for me, he wanted to come up and chat. I had noticed that Mark's car was not in his garage space, when I parked. It seemed obvious to me that Jordan knew that, too. He came over within the hour. Even though he had been courteous, polite and friendly, due to my history I was not sure about his coming into my apartment. My apartment reflects my interest in ufology, even to the extent that I have a space craft and a star designed into the tile work on my floor. I did not want him to dismiss the importance of our problem with Mark, due to what he might construe as my idiosyncratic interests.

Jordan kept his appointment right on time. He immediately commented on my wonderful ocean view and the apartment in general. I had shelves filled with UFO books. He started to browse through them and asked if he could borrow a couple. I showed him a few photos of ufological phenomena that I had on my walls. He looked with interest. I offered him a soft drink and when he spotted the tile work spacecraft, he exclaimed with delight how "neat" it was. He jokingly asked if my tile man would do a FBI seal on his kitchen floor.

We had more pertinent discussion about Mark. The bottom line was that he had not done anything illegal...yet. The Bureau was concerned and wanted to be in touch with the situation. He asked me to keep him apprised. It was a very pleasant visit.

After he left, I thought how amazing my life was. It was not my intention to make the issue of Mark anything other than it was with the FBI. Somehow, even though I pointed myself in one direction, the path led back to the issue of UFOs and aliens. Was this another incredible coincidence or was this a contrived situation? I felt that Jordan was quite sincere. I asked him again if I had a file with the Bureau and he assured me that I did not. I felt no fear of this.

How the real life "X-File Man" found me, I don't know. Knowing the level of coincidence in my life, the answer could only be moot. My life was now getting *really* interesting, *but I had no clue just how interesting and truly bizarre this scenario was going to become...*

INTELLIGENCE AGENTS:
AN OXYMORON?

For several years prior to the appearance of Jordan, the "X-File Man," I had kept myself celibate. All potentially romantic male relationships were kept at arm's length. On a soul level, it was to keep me pristine, focused solely on my path and the tasks ahead. On the pragmatic, mundane level, I did not want to share my life with anyone who was a UFO skeptic or a disbeliever. I did not have the time or patience for explanations or justifications about following my bliss in UFO research. I did not want anyone jealous of the time spent on my passion and my thirst for more knowledge. I also was accustomed to travelling at a moment's notice in pursuit of this knowledge. This large amount of travel would not bode well for a tight relationship. I had more than my share of romantic relationships in the past, so I decided that the path of least resistance was to eliminate them completely until my work was finished. It was not so easy to break such a life long habit, but with determination it was done.

I found that celibacy in itself was very empowering, as a lot more energy was left to live life and most of all, be centered and relatively free of soap opera anxieties. As far as I was concerned, like nuns who are brides of God, I had become the Bride of the Cosmos! My cosmological inquisition provided a very happy, fulfilling relationship and as the years passed, I saw no reason to alter the status quo.

I was rigid and inflexible about this. My friends shook their heads in wonder at my hard core stance. My resolve was unshakable... until the "X-File Man" entered my life. I must admit that he did knock me a bit off my immovable objectivity. What he was in my life and how direct his purpose was as an FBI Special Agent is still unknown. I wish I could know for sure, but I doubt if I ever will!

One of his most titillating revelations was that he had been in Naval Intelligence after he graduated from Duke University. He then worked for the CIA for a few years but decided that he didn't want to be a "spook", so he applied to the FBI and was very happy there. What an intelligence troika that is! For a UFO investigator this could be the most terrifying situation. For me, there was an exciting, perversely exhilarating challenge in knowing and being with him. My fearlessness was a facet of my personality I had not been acquainted with before.

Jordan Perez's demeanor to me was silky smooth. He was incredibly well mannered and a perfect gentleman. He acted as if he had been through a military academy and a finishing school combined. Soft spoken and non-aggressive, he was most women's dream man. Even though he was average looking physically, the characteristic that most appealed to me was his mind. He was very articulate and knew how to handle a good conversation using the English language well. I was never bored with Jordan

Although most of our conversations revolved around ufology, surveillance and his personal problems, he was also extremely well-read, a wine connoisseur and a gourmet. The guy just knew a lot about a lot of things and I enjoyed being with him. I was impressed with his intelligence and charm. This plus the fact that he was an FBI Special Agent, theoretically a macho supercop, possibly surveilling me, kindled something perverse in my nature. This was someone I was supposed to be fearful of, probably the "enemy," but I wasn't afraid. Despite that, *I never* stopped analyzing his actions and words that might betray his true focus in pursuing this friendship. Somehow, that was part of the fun!

After the initial interview with Jordan at the restaurant and his visit the next day to my apartment, the "relationship" escalated at Jordan's insistence. For all intents and purposes, it was for his personal growth and knowledge about ufology. We discussed and shared books and articles. He would tell me about unclassified information he was getting from the FBI computer data base. He evidenced a real interest and amazement at all of the new knowledge he was amassing, of course, with my help! He even said that he was considering going back and reevaluating cases related to MIB's and other anomalous interdimensionalities that he had construed before and reported on as bizarre and unbelievable.

Tangentially, we discussed Mark, the condomaniac, and his continuing unpleasant activities. Jordan said that he was still working on the case and hadn't made a final determination or report. He was no longer concerned about Mark's seeing him.

At first, Jordan phoned me daily to chat. It quickly escalated to several times a day and also late at night from his home. I was friendly but tried to be detached. He was getting what I call "the voice." That is when a man becomes more than superficially interested in a woman. "The voice" sends out an energy pattern verbally that is like a siren's warning blast. Jordan's "voice" was unmistakable. He wanted to move the friendship into higher realms romantically. Normally, this would not be a difficult situation for me to deal with but there was even a more insidious situation than his FBI status; the guy was married! He wore a wedding band and never spoke about his wife or family. I know how secretive police and agents are about their families, mostly for security reasons. I did not want to breech that and never asked about his family, nor did he ask about mine. I felt it was now time for me to push the envelope. I had to, because I did not want to get involved with a married man! It has always been my moral imperative not to support men in illicit, adulterous relationships. I was a wife once, myself, and have never forgotten that! I had to tread lightly because he had not completed the report on crazy Mark and I wanted his support. That was what this was all supposed to be about anyway! This was going to take some diplomacy!

My opportunity to finesse the situation quickly presented itself. Jordan knew that it was my birthday. He suggested that instead of meeting for lunch, as we often did, he suggested that we meet at an upscale Ft. Lauderdale intracoastal restaurant for dinner. He said that we would have more time to talk, since he has limited time to spend at lunch. In this case, I was fearing that "talk" was meant to be a euphemism, since it was a very romantic restaurant.

I valeted my car and waited for Jordan to appear. I was nervous, trying to decide how I was going to handle him diplomatically. He soon appeared, put his arm around my waist and kissed me lightly on the cheek. He was very complimentary about my appearance. I now felt sure about his interest in me.

The meal and wine were wonderful. When he offered me a taste of his entree, from his own fork, I knew he was really serious. I remembered reading that when a person of the opposite sex really wants you, they will offer you food using their own utensil. So, I guess, all of the signs were there. It was now or never! I changed the subject we were on and asked him where he lived. He replied that it was in Kendall, a neighborhood in Miami, at least an hour from Ft. Lauderdale. Then came the biggy! "How many kids do you have?" He looked down at his plate and said very quietly, "three." "How old are they?" I pressed on. Still not looking up, he replied, "Eight, seven and three months." In my mind, I was fuming. I thought to myself, "How typical. With a three-month-old, he can't say he isn't getting any!" I continued to inquire, "Are they boys or girls?"

"The oldest is a boy and the others are girls." He still wasn't looking up and was playing with his food now. And in a barely audible voice Jordan said, " but my wife and I are having problems. We are not getting along." "My God," my inner dialog continued, "what a trite cliche. I would have hoped for something a bit more creative!" I didn't respond to Jordan about his personal problems, as I really didn't want to soothe his "wounds" any.

We finished dinner and went for our cars. Jordan did not valet his. He uses various FBI cars and never valets. I guess it is some kind of a rule or maybe his gun was in it. I decided it was my opportunity for a quick get away. Jordan came back to my car with some books he had borrowed. "I really enjoyed talking and being with you, Mia. I hate for the evening to end. I know that I can't go back to your place, since Mark might see me and would recognize me. Isn't there any other place we can go?"

Without hesitation, I said, "No. Thanks for dinner. Bye." As I drove back to my apartment, I breathed a sigh of relief. I hoped that he wasn't angry but no way would I get involved with a married man, not even if he *was* the X-File man!"

Jordan, however, was not deterred from his course. The next morning, he phoned me from his office, thanking me for joining him for dinner the night before. "Ever the gentleman," I thought, "even when he is rebuffed!

"I have something to tell you, Mia. Something that I have not told anyone, even my marriage counselor. When I told you last night that my wife and I don't get along, it is for a very serious reason. The three-month-old baby I mentioned is not my child. My wife had an affair and I gave the baby my name. Why should an innocent child suffer? She is a beautiful baby but she doesn't look at all like either of us. She has very fair skin and big blue eyes. My wife and I are both of Cuban descent and have dark eyes and hair. Neither of us have blue eyes in the family. Every time, I look at the baby, I am filled with rage at my wife's deception and betrayal. When she told me she was pregnant, I was amazed, since I had a vasectomy after my daughter was born seven years ago. I went to the doctor and he told me that I was still shooting blanks, so I knew for sure that the baby was not mine. I am totally sick about this and hate my wife. I can't stand to be anywhere near her. My children have noticed this and are upset. I love them very much and don't know what to do. I have filed for a separation but can't stand the thought of leaving my kids. They are great kids and I am very close to them. They depend on me to help them with their homework and coach their teams. I was lost until I met you. I am falling in love with you and now I have the courage to leave. I am moving into a motel. I am getting the corporate rate, so it is more affordable. My godfather, who is only 17 years older than I, said that I can move into his home. He and his wife are teachers and don't have any kids. They are busy people and said I wouldn't get in their way."

That was a lot for me to digest. I was bowled over at his revelations. I had never heard a man express himself like that. He sounded so emotional and pitiful. I thought he was going to cry. "If you need someone to talk to, you can count on me," I said. I really didn't know what to say. I didn't want to get anywhere near that "I am falling in love with you" statement. I might have been attracted but I wouldn't allow myself to "love" any married man, even though now he was talking about separation.

The bit about his wife's affair and pregnancy was a shocking revelation from any man, let alone an FBI agent and a Latino! I was amazed that he would tell me that. I had only known him for a month! I certainly never expected my problems with condomaniac Mark to lead me into a whirlwind romance with a cuckolded G-Man!

Jordan's time out of his family home was short-lived. He did give me the phone numbers of the motel and his godfather's home. So, I assumed that was the real reason for his being there. This all lasted less than three weeks. It was around the Christmas holidays. I warned him how difficult it would be to stay away from his family then, and that his leaving was ill-timed. My prediction was correct. He soon moved back into what he said was a separate bedroom from his wife. I felt no press about this. It gave me a security blanket to keep him from escalating the scenario.

Whenever he accelerated his romantic overtures into mentioning lovemaking, I repeated my moral imperative about married men and especially those who lived at home! Jordan never gave me any problems about that, and, as a matter of fact, he said he agreed with me. He stated that if his children were to find out, they would be crushed. This really seemed strange to me. I never knew a man who had been without sex for a long time, to care what his seven and eight years old kids would think about his sexual escapades. It seemed totally incongruous and unnatural, even for this perfect gentleman from the FBI. Even J. Edgar Hoover had his peccadillos!

As this totally unexpected liaison was unfolding between Jordan and me, I remembered something that I had said like a mantra for a few years since I was aware of surveillance by an agency or agencies. I didn't think that I was being surveilled so much in the last year, so I put it out of my conscious mind. I had mentioned to several friends and also to my Excelta contacts that I had a very strong and omnipresent feeling that one of my "watchers" was totally intrigued by me, so much so that he was infatuated. The "words on my window" after I wrote the investigation on The Professor and The Professor seemed to be proof of some kind of an attraction. Could impetuous Jordan be the guy? Was this a set-up? Where would this lead?

Outside of the area of romance, lots of other interesting things were percolating. Jordan continually excoriated my good buddy, Andy, the private detective. Even though Andy was the catalyst for getting us together, there was nothing about Andy Jordan liked. It appeared that he went out of his way to do an investigation on Andy. He said that everything Andy had told me was a lie. Jordan then offered to help me get all of the *correct* information from computer data banks that Andy had lied about.

I knew that Andy had not always been "totally accurate" about his work, but he did have a way of getting information that was proven to be accurate and that other private investigators wouldn't handle. Perhaps, Jordan was trying to cut me off from Andy by defaming him and putting negative thoughts into my mind regarding Andy. Jordan had mentioned early on in our friendship that he had this very strong feeling that he had to be my "protector." He felt that it was his job to take care of and protect me. I found that a really interesting statement, since what woman alone doesn't, at some time, wish that she had a protector to be her Sir Lancelot. Was it possible, that I had mentioned this desire on the phone or elsewhere, and that statement was being used to rope me into the net even tighter. Or were we under some kind of mind control scenario where both Jordan and I were programmed to feel that we were destined to be together?

Another good argument for possible mind control links were the synchronicities that occurred so often. When Jordan said that he was "falling in love" with me, I mentioned to a friend that Jordan was becoming "smitten." It is not the average expression in the 90's to use but it is the kind of word I would use. In a conversation with me about the genesis of his feelings, he said, "When I first met you, I was attracted but soon after that I was smitten!" I couldn't believe my ears. I found it really incredible for a 32-year-old guy to say that. Either we were soul mates or he was listening in on my telephone conversations!

As we entered into the second month of this intensive alliance, Jordan seemed to get very caught up in his focus on ufology and of course, helping and protecting me. He said his job was suffering because of this but that he didn't care. It was at this time that Jordan began to remember two possible alien abductions he had experienced. One, he said, occurred when he was six years old and lived in Queens, New York and was taken from his bedroom. The other occurred while he was jogging on a course, while attending Duke University. He felt he had missing time then, for sure. He told me that his marriage counselor had given him some psychological tests that showed that he had post traumatic stress syndrome, and recommended that he go through hypnotic regression to find the source of the trauma. Since he had not been in a war or had any other trauma that he knew of, other than his wife's infidelity, he was perplexed about this. It seemed a convenient conclusion for me to suggest that perhaps he had repressed memories of alien abductions. The serendipity of the stress syndrome and his two memories of possible contact seemed contrived. I was starting to tally a mental ledger of debits and credits on Jordan's credibility. Lots of facts and happenings of high strangeness were being heaped on me. The gamut of ufological booga-booga seemed to be appearing from a cornucopia of orchestration. Everything I had ever read, and some I had never imagined were becoming "normal" events. I wondered who was masterminding these plots and for what reason was I being spotlighted?

Jordan said that he was surveilled by a male in a car that had untraceable license plates while he, himself, was doing surveillance on a case. He also said that he and his family were followed to church one Sunday by a woman in a van with Alabama plates. I suggested that perhaps she was from NASA since they have a facility in Huntsville. Jordan also showed me a relay type of listening device that he said he found in the telephone circuit box outside of his home. He claimed that he found large male footprints leading to that area, in newly laid sod.

The most fascinating of Jordan's renditions of being watched happened when he was in Atlanta for a two-day FBI training session. He and his partner, Roger, have similar appearances since they both are of average height, have mustaches and wear glasses. Jordan said that before he left for Atlanta, he decided to shave his moustache and wear his contact lenses. His partner had reserved both of their Peachtree Plaza hotel rooms in Roger's name. On the first day of their training, at the afternoon break, they each went back to their respective rooms to freshen up. Both Jordan and Roger said that their belongings had been moved, as had the furniture. This was done, *after* the maid had cleaned. The next day, both Roger and Jordan woke up with strange red marks, and rash like welts, across their backs. Roger had an intense headache and told Jordan that he felt that someone had been walking around in his room, while he slept. Roger's headache continued after he returned to Miami and his dreams of an intruder(s) had intensified into nightmares. Jordan hypothesized that the intruder(s) were really there for him but since he had changed his appearance, Roger fit the description better.

Thus Roger had reaped the wrath of the watchers. Before Jordan left for Atlanta, he had asked if I would meet him up there, since he had his own room. Of course, I declined. Perhaps, if I had gone, something would have happened to me for which I was not prepared!

Synchronistically on the day of Jordan's return from Atlanta, I had very strange tones left on my answering machine. It was a combination of high frequency pitches, screeches, tones, beeps and a super-high speed voice underneath it all, a real mixed bag of sounds. This was the second time that had happened. The first time, I answered the phone and heard it in person, and it happened one more time subsequently. I came home in the late afternoon, checked my caller ID and played the answering machine. It was shocking to hear those unsettling tones again. I was wishing that Jordan was back to hear it, but I didn't know when he was going to return.

As soon as I had that thought, the phone rang. It was Jordan calling from the Miami airport. I told him that I was so glad to hear from him and was just thinking about him because I wanted him to hear these strange sounds on my answering machine. I asked him why he thought to call me now? Jordan's response surprised me. He said that he had just arrived and was walking towards the baggage carousel, when he turned on his beeper and it went off with my number on it! I told him that I had *not* beeped him and that I, myself, had just come home. I had no idea when he would arrive. He agreed, since he, himself hadn't planned on being back so soon, but had taken an earlier flight. This situation certainly offered a clear example of how the watchers were on the job trying to confuse, confound, perplex and intimidate. In my gut, I knew that even weirder stuff was en route.

After Jordan's FBI training days in Atlanta, everything accelerated into a fevered pitch. His ardor towards me was constant. Phone calls, day and night, filled with expressions of his love and concern for me, always wanting to know if I was all right. Perhaps, he was just fishing to see how I was responding to all of the stimuli heaped in my direction.

Most potently insidious was his revelation to me about the man who got his wife pregnant. He informed me that Homestead Air Force Base, which is south of Miami and was devastated by Hurricane Andrew a few years ago, was now home to a state of the art technical facility of the Air Force Office of Special Investigation. The AFOSI is the intelligence arm of the Air Force. Jordan stated that the focus of this facility was to gather information on UFOs and underwater anomalous activity most likely connected with UFOs. Jordan said that the man who impregnated his wife was an AFOSI officer stationed at Homestead Air Force Base, who was also married and had a child. I asked where she had met this guy. This is when the information turned beyond the pale of high strangeness.

Jordan said that he and his wife Joanne, had been invited to a party given by his immediate boss, Mike, for FBI Special Agents and their wives. This AFOSI officer appeared without his wife and in full dress uniform. He was the only non FBI person there. It was at that party that a romantic connection was made between Joanne and the AFOSI guy. Jordan said that it lasted at least one school term because Joanne was going to a night school class but received an incomplete grade at the end of the term, even though she supposedly attended each week! I asked who invited the AFOSI man to the party. Jordan said he asked his boss and his boss said he didn't know. The fellow just sort of appeared there. This was too unbelievable. The party was held far away from the base in another county.

Jordan then supposed that perhaps this was not just happenstance, that this was an orchestration by the powers that be, to get him to leave his wife and be with me! He said that all of the time he was away in the Navy, his wife never cheated on him. So, why now? I asked what happened after Joanne revealed the true father of her child. Jordan said that he was enraged and told her he was going to kill the officer. He made her tell him where he lived, then drove in a frenzy over to his house. When Jordan arrived, he found the house closed up and a Century 21 "For Sale" sign on the front lawn. When he checked up on the officer's whereabouts, Jordan was told that he had been transferred to a Pentagon-related agency.

I just sighed when I heard this. Was I supposed to believe this science fiction scenario? It was a bit too much, although Jordan was acting as though it was Gospel. Was he saying this for me to believe and think that somehow Jordan and I were to be considered a unit, and that this had been done to get us together?

Apparently, the AFOSI officer was the Judas Goat that led Joanne astray and the two of us were to follow who knows where! My UFO researcher friends that lived near Homestead Air Force Base had checked their sources there at the base and said that there was no AFOSI technical unit there. These were tricky guys, these intelligence agents! One thing I *could* say for sure was that I didn't believe *any* of that story. It was an unbelievable crock! Jordan was certainly underestimating me and overestimating my gullibility.

I did not acknowledge my disbelief. I was caught up in this chess match and I was fascinated to see where the strategy would lead. In my heart, my greatest hope was that somehow "they" would tell me why so much time and energy has been focused on me. Why was I so important and what did they want? I was hoping that somehow and in some way Jordan would let me know. Many times I made a statement to Jordan that even if he *was* my enemy and I was his assignment, it didn't matter because we were bonded by something greater, the Laws of Karma. I told him that I would love him unconditionally, nonetheless. He never rebutted the statement of his culpability. He did latch on to the bonded relationship idea. He probably believed it, and maybe in a strange way it was true. Whatever integrity Jordan possessed prevented him from saying that he wasn't the pursuer and that I wasn't the pursued. Maybe that was some kind of weird FBI code of ethics at work. So far, it was the *only* ethical stuff I had seen!

During Jordan's siege with FBI data base on UFOs he said that he was inundated with information on the subject. One day, he found on his desk a file that supposedly was marked "ABOVE TOP SECRET". This was highly irregular and an obvious glitch. He was not classified high enough to be in receipt of this kind of information. He said that he opened it anyway. Before he could continue, I told him that I did *not* want to know anything about it. I was going to hang up the phone if he continued. This felt to me, like a set-up. The next day, he told me that he had read the file and it was extraordinary but he felt compelled to tell his immediate superior what he had and what he did. Eventually, there was a meeting with his superiors about this and he was cleared of any disciplinary actions.

The other reading material that just happened to cross his desk, unasked for, was Phil Klass's latest UFO debunking book. He asked me if I knew about it and Klass. I assured him, I did. I told him that I had even been on a local radio program that was debating the UFO issue. Phil Klass was talking via the phone and asked the talk show host specifically what my name was. Eventually, Klass hung up, after being harassed by the host. I really didn't get a chance to say much. Jordan said that the book came quite mysteriously in an envelope from Washington, DC. There was no return address, even though a U.S. government postage meter was used to send it. Another curious thing about the book was that even though the bindings looked fresh and appeared to not have been broken, as if it had never been read, there were parts of the book that were highlighted. These parts mentioned the FBI and alien abduction hypnotic regression fallacies. Just another coincidence in the chain of coincidences. Now what was I supposed to infer from that?

For the following two weeks, I was besieged by continuous telephone call and caller oddities. I could mentally see the creators of this malicious mischief sitting in their little offices thinking up these antics and rubbing their hands with glee. To be able to accomplish most of these telephonic extravaganzas takes a state of the art level of scientific wherewithal and expertise. Here is a retrospective of some of the high tech happenings on my phone.

One night just before midnight, I was sound asleep. The phone rang and a woman who sounded as if she had a Haitian accent asked if I was the one who had phoned her an hour before. Sleepily, I told her that I hadn't. She asked if there was anyone else who lived with me who could have called. I told her I lived alone. She insisted that my number was on her call return and that she just punched the return call button and got my number. When she read out the number that was on the display, it had one number that was different than mine. I told her that the number was almost the same but different. She said that was impossible, because the call return was computerized. I agreed that was strange. I asked her what her name was and she said "Mia." She asked my name and I told her it was also "Mia." We both were surprised at that. I was especially intrigued, because there are not many, if any, Haitian women named Mia. How did a call return call with the wrong number send a computer signal to my phone and coincidentally have the caller and the called have the same unusual first names?

Three times on different days and at different hours, a strange mechanical voice appeared on my line. As soon as I would pick up the phone, and without touching any numbers on the key pad, I heard this message: "WE ARE SORRY, THERE ARE NO ENTRIES ON YOUR LIST FOR HELP IN TURNING ON THIS SERVICE. PLEASE HANG UP NOW AND CONSULT YOUR WRITTEN INSTRUCTIONS. THANK YOU." I had not received any calls nor made any calls just prior to these recordings. I had not received any advertising automated messages or talked to anyone with a voice mail, nor do I have voice mail or a fax line. I checked with the phone company and they were baffled. A friend who is familiar with intelligence agencies systems stated that this probably occurred due to the fact that there were at least two different surveillance taps on my phone, probably from separate sources. If one was not aware of the other and fed in a new surveillance line, it could send a signal to the other that would trigger a kind of a "feed back" or glitch. So far, this was the only explanation that had come forth.

In the last few years, since I have had many strange phone calls that needed to be documented and saved, I have devised a technique that has helped me get the job done. I have done this simply and inexpensively, with the technology available to the average citizen. Mostly everyone in the civilized world has an answering machine now. God forbid, we should miss a call! I have one that allows for a one-minute incoming message. I also have Caller I.D., which is even more indispensible to me. The only calls that can't be identified are those outside of the local calling area and ones from cellular phones. These calls are identified as "Out of Area." All of my weird calls had this notation on my Caller I.D. I also attached a separate cassette tape recorder to my phone line which picks up all incoming and outgoing messages, including those on the answering machine. It is interesting that the tape recorder is so sensitive that it has picked up tones, beeps and editing of my calls that I would not have known about, if I just listened to my answering machine messages. I did not like to have to take these measures of security but I knew that it had to be done. I have always made it a point to remember to have my tape recorder on, with enough tape available to handle unusual phone activity. It has absolutely proven to be a great chronicler of the insidious activities of others, which will be described later in this chapter.

The day after Jordan revealed the details of who his wife's lover was and how they met, I had an astounding phone call. I was at my desk busily reconciling my bank statement when the phone rang. "Is this 305-876-9407?" the caller asked. I hesitantly said that it was. He went on, "This is Chet Olson of the Air Force Office of Special Investigation. Did you beep me?" I was a little taken aback. "Is this some kind of a joke?" I asked hopefully. Olson laughed heartily,

"No, it's not." I took a deep breath and tried to think quickly on my feet. If he was the real thing what should I do? "Did you say you were from the A.F.O.S.I.?" "Yes," he replied, "didn't you beep me?" "No, I didn't," I replied emphatically.

Olson again repeated my phone number and asked if it was correct. He seemed confused. He said that my number had just come up on his beeper. "Where are you calling me from?" I asked. He replied that he was calling me from a cell phone in Tampa. I remembered that Tampa was the home of MacDill Air Force Base, hypothetically one of the bases that houses UFO hardware and alien bits and pieces. Still sounding confused Olson said he couldn't understand how my number got on his beeper. He said that he was going to check his office to find out since his beeper number was secured.

"I'll be happy to talk with you, anyway," I said cheerfully. He laughed again, saying that he was sorry that he bothered me, then hung up. In a matter of seconds my phone rang again. It was Olson. He was now thoroughly confused, saying that he was having a difficult time getting my phone number off of his cell phone's memory. He again apologized and hung up.

Now that conversation was really one for the UFO investigator's handbooks! Olson sounded very legitimate or he was a very good actor. I really wanted to talk to him some more. I would have loved to tell him who I was and what I was interested in and chat. But I later thought that it probably was better to stay anonymous. Although, if he really was who he said he was, I am sure he had checked me out and knew everything about me by that time, including whether I flossed my teeth or not! How *did* my phone number get on his beeper? Was this another move on the chess board to intimidate me? I was beginning to have a reverse reaction to these efforts. I was beginning to feel really proud. Maybe I *am* really somebody special? Maybe I *do* know something outside the ordinary UFO crapola? Maybe I *will* be doing something that "They" want to know about, sometime down the line? My mind boggled with the possibilities. At that point in time, I thought "They" were wasting a lot of theirs.

When I told Jordan about this phone call from Chet Olson and played the tape of the conversation for him, I asked if it sounded like his wife's former lover. He said it did not. Eventually, Jordan said that he checked up on Olson and that he was the top A.F.O.S.I. guy at the base and that Olson's beeper number would be on a secured line and would be secured information. I had another corroboration that Olson was, indeed, a legitimate A.F.O.S.I. officer by an employee at MacDill Air Force Base. Not only does Olson exist, but he is the District Director and top intelligence officer for the entire eastern region, and maintains an office at MacDill. In further discussion with other A.F.O.S.I. personnel on base, the MacDill insider was informed that the Air Force, despite their protests to the contrary, currently *does investigate* UFO sightings. Whoever got my number put on Olson's beeper must have had some major connections... big connections... but then again maybe it wasn't *really* big shot Olson that phoned. How would I even find out without bringing additional attention to myself? Was this all Jordan's doing? This continued to be a real hall of mirrors.

The day after the A.F.O.S.I. man called, I had an equally fascinating conversation with Greg Warren of the Florida UFO Hotline in Pensacola. I was again at my desk in the afternoon trying to get organized, when the phone rang. It was almost 24 hours to the minute from Olson's call. The hot line phone is a number to call, anonymously or not, to give information related to sightings, abductions, etc. They can also ask for information on that phone number.

I had not met Greg but I was familiar with his name as a researcher. Greg's question to me was becoming a familiar refrain. It was practically the same scenario as the Haitian "call return lady" and Chet Olson's.

"Did you call the UFO Hot Line last night?" Greg inquired. "No, I didn't. Why are you asking?" Greg answered my question with another bizarre response. I was now becoming used to this craziness. "Last night, a very strange message was left on the hot line by an anonymous, disguised voice. It was a purposefully "coded" voice that was recorded at high speed and then played back at low speed. Even though it was made to sound like a man's voice, I have a gut feeling that it was really a woman's. I played the message over 100 times, until I got everything it said."

"Well, what did it say?" I asked impatiently.

"It said, 'Someone needs to look at the Golden Sands Condominium in Ft. Lauderdale, 12th floor. Regular abductions visible from the outside, large craft, late at night.'." I shook my head and closed my eyes in absolute wonder at this latest episode. I *do* live on the 12th floor of the Golden Sands Condo in Ft. Lauderdale. "Did they say who was being abducted?" Greg then went into a long discussion as to how he deduced it was me the message was about. "No, they did not say who it was but it didn't take me long to figure it out. I have a computer program that has all of the phone numbers in the United States. I found Ft. Lauderdale and then the Golden Sands and then the 12th floor and got the names. There are only eight apartments on that floor. I thought that it was someone who was on the hotline newsletter mailing list because they would know the hotline number. Since you are, I matched your name up to the condo phone owners. I almost thought that it was a test to see how good of an investigator I am," he said proudly.

"I must congratulate you. You did a great job and are a good investigator, but I think that there are a lot of others who have the hotline number who are not newsletter subscribers! Even though there are eight apartments on this floor, I am the only one who is here full time. The rest of the owners come and go or are "snowbirds" and come solely in the season. So, whoever called the hotline had to be talking about me. I have had a lot of surveillance lately and this jibes with the rest of the activities. As far as my being regularly abducted, if I am, I have no awareness of it. I sincerely doubt that anything like that is happening to me, but who knows?" I responded.

"Well, it sounds like some kind of a prank," Greg declared.

"I don't have any adversarial relationships that would merit someone going to that much trouble to try to embarrass me or put me on the hot seat," I told him. "No, this sounds like government types, just creating harassment and continuing to make a nuisance of themselves."

"What are you doing that would make them so interested in you that they would do this?" he asked.

"I wish I knew, I wish I knew! I am not doing anything other than just communicating with other researcher friends. Last winter, I wrote a case that was included in a Leo Starr monograph on crash retrievals but I am not working on anything else now."

Greg then continued to give details of the recorded message itself and the technology needed to record the obscured message. A few weeks after I spoke with Greg, the hotline newsletter arrived at my home. Within the newsletter was a small blurb asking for the person who called the hotline from Ft. Lauderdale, the day the coded message about me was left. It said to please contact the hotline numbers again. The blurb went on to say, "We have to talk." I guess Greg didn't believe me!

Eventually, when Jordan phoned me, we entered into a new *non* phone message scenario. Despite the fact that I have a fairly long time unit for incoming messages on my answering machine, whenever he would call and leave messages, they would be cut off after a few words. He had no idea this was occurring and would continue talking. When others would leave messages before or after

Jordan's, the messages would be full and complete. When I checked my tape recorder, it had recorded Jordan's *full* discussion including a signal to end the answering machine message, even though he was still talking. At the end of the appropriate time unit, the real tone signal sounded. I then heard what sounded like a cassette tape whirring. This was most likely the sound of the incoming message being edited by unknown hands.

One time, my Caller ID showed that I had received an Out of Area call, but there was no message at all on my answering machine. When I played my tape recorder, it revealed Jordan's full message that had been left on my answering machine, including my outgoing message and the normal tones at the beginning and end of the message! Therefore, it seemed that "they" were erasing his messages partially and totally. If my answering machine failed to operate properly, it was only when Jordan called. If I hadn't had the hunch to check my cassette recorder, I would never have known that Jordan had left a message at all on my answering machine, since *it had been edited out completely!* These agents were busy little bees around me. They probably had the code on my answering machine to erase messages, or else they came in person to my home and did it manually! The thought of that really ticked me off!

When I mentioned the supposition that these guys could possibly have come into my home and have done that, Jordan gave me a very frightening suggestion. He said that if I ever inadvertently encountered someone I thought might be an agent in my home, I should just look away from his face, if not down at the floor and let him get away. "Don't even acknowledge him or threaten him!", he added. He further declared that if that occurred, I should throw all of the food in my refrigerator out, since "they" might have poisoned or contaminated it. These are the words of a federal agent! Now doesn't that send a chill down good citizens' spines?

To further complicate the communications situation, Jordan told me that he had not received several of the phone messages I had left on his voice mail, as well as calls I left on his beeper with my phone number. Was this more manipulation or the truth? Did he want me to think that there were also agents checking him out and wanting to keep us apart? Maybe there were. He also said that he had two strange messages left on his voice mail. One had the sexy, voluptuous music of Sade, with the sounds of people moaning in the backgrounds. The other message was left late at night. It was just the sounds of a busy office, with typewriters, machines and the drone of voices. It seemed inappropriate for heavy duty office sounds at a time when most offices would be closed. There weren't any verbal messages left in either recording.

Despite Jordan's continuous devotion to pursuing a romantic relationship, I began to feel that his interest in helping me solve some of the riddles about my past and present surveillance was just a red herring. It seemed that so much was being done to encourage my absolute faith and trust in him. He wanted me to think of him as a kindred spirit. He even went so far as to have his eight-year-old son talk to me on the phone. Jordan said that his son was going around the house tormenting his younger sister with the threat that he was going to have her "abducted by aliens!" According to Jordan, this seemed to have come out of nowhere, and had never been done before by his son. He had the boy tell me exactly what he had told his sister. Now, Jordan wondered if his children and wife were also involved in this intrigue? He certainly wanted me to think that the thought horrified him.

Several times I mentioned my ufologist mentor and greatly admired friend, Leo Starr, an internationally renowned researcher who had recently died. I thought that perhaps my friendship and working relationship with him might have catalyzed to the maximum an already obvious interest in me by "Them." Jordan took it upon himself to check Starr out by asking for his "file." He told me that Starr's file was several inches thick and had "Secret" and Above Secret" classifications on it.

Then, in an obvious attempt to discredit Leo in my eyes, Jordan said that there was some question of a "foreign flag" involved with Starr. Jordan shrugged his shoulders and said this with a sneer. "Are you saying that he was unpatriotic and aligned himself with a foreign country?" He gave me another shrug in response to that. "I don't believe it.", I protested. "Leo loved America and was a great family man. He didn't have a profession where he was near anything that could compromise this country. I know that he did have a prince from a tiny European country stay overnight at his home. This prince is very interested in ufology and is a benefactor to researchers who have special research projects. That is about as 'foreign flaggy' as Leo ever got. Leo mentioned this fact to me, because when I visited his home, I slept in the same bedroom as the prince. It was mentioned to me jokingly."

Jordan then went on to say that I should be careful to not align myself with a "foreign flag." I said, "Do you mean, if an extraterrestrial says hello to me, I should run the other way?" Jordan did not answer.

This issue of Leo Starr, I believe, was some kind of bone of contention for Jordan, probably because I admired Leo so much. I had him read my report and case investigation that was published in Leo's monograph. Jordan made absolutely no remarks about it being a very remarkable and unusual case. Leo had even mentioned to me that my case investigation had gotten more comments and feedback than any of the others in the monograph. Jordan couldn't care less. He didn't even do the politic thing and comment on how well it was written or that it was even interesting.

"Very peculiar," I thought, "very peculiar. Jordan bears some real watching." I was now scrutinizing every word and thing he did. That was a good thing, too, because he kept making a lot of mistakes. On my mental ledger, his deficits were beginning to outweigh his assets, in a major way! But I still wanted to keep in the game. I had to see where this was all heading. Knowledge is power, and I felt that my knowledge that this relationship with Jordan was not what he wanted me to think gave me an edge. At least, I prayed that it did!

One evening, Jordan arrived to take me to dinner. We had some wine and pleasant conversation He got up and started to walk around my apartment. "I should really come over with my equipment and check your apartment out for listening devices. These "bugs" are so small now, they can fit even in the tiniest places." I told him that after all I had been through, I assumed that they already were listening in and they must have been getting pretty bored. I knew that with the current technology, all they needed was my phone and my television to bug my apartment. I had heard that with microwave technology, all that was needed was pretty much just an arrow pointed in the right direction in order to focus in on their prey! It had been years already since I had become their project, and nothing that they would be interested in, or that I *think* they would be interested in, had happened. Despite my deep concern for the violation of my constitutional rights of privacy, I had gotten used to "them."

My apparent lack of concern did not deter Jordan. Even though we were almost out of the door, heading to the elevator, he turned and started to look around my apartment. All of a sudden, as if he got a great flash of intuition, he pointed towards my smoke detector, in the hallway leading to my bedroom. He asked if I had a step stool. I retrieved one from the closet. He mounted it and took off the covering of the detector. In a whisper, he asked me if I had a small screw driver. I nodded and went to get it, out of his vision. In a few seconds, when I returned, he put his finger to his lips and motioned for me to be silent. In his other hand, he had a tiny little black plastic thing. He handed it to me and put the cover of the smoke detector back in its place. He returned the stool to the closet and told me that the little black plastic thing was a high tech listening device.

Immediately, I was on guard. I did not like the fact that I did not actually see him pulling that device out of my smoke detector. It had tiny little prongs that looked like it should fit into something else. I asked him how this was connected to something electrical, since it had to have wires or another part to it. In a very off handed and generally non specific statement, Jordan said that what it was connected to had fallen back somewhere into the innards of my smoke detector. I decided not to take this to an intensive probe, until we returned from dinner.

When we did, I requested he show me where the wires or connections were. I was still not satisfied. I got on the step stool and saw that this device could not have had anything else connected to it. Jordan again tried to finesse the situation with double-talk.

The next day I phoned my good buddy Harvey, who is an electrician and is also a UFO buff. I showed him the device. He took apart the smoke detector and could not find any "female plug," which was the gobbledygook answer of Jordan's. I then went to the local "Spy Shop" that sells surveillance and protective devices, and asked about the hard little black plastic listening device.

Mr. Brown, the shop's owner, was very knowledgeable in this field. I showed him the device and asked if he had any of these in stock. He took it from my hand and smiled. "I have to order this brand. We don't normally keep it in stock. It is an excellent Panasonic device. It picks up all conversation within 40 feet. Of course, you know it has another part that fits onto these prongs. It won't work without it." I asked him what it cost. "With the other part that it fits into, the price is $555.00. I thanked him and left.

Jordan was playing another game with me. Again, he underestimated my confidence in him. Why did he think that I would be so stupid as to accept a non-answer from him? Here was another mistake to add to his growing list of inscrutables. Since this was, indeed, a real listening device with an additional and integral part missing, the most rational assumption to make was that he took the device from the FBI's stock pile of devices and tried to palm it off as coming from inside my smoke detector. I am sure he thought that this would be the perfect way to ingratiate himself, in my eyes, to help save me from the unseen forces.

He attempted to make me feel that he was, or might be, a fellow victim or subject of manipulation by the Powers that Be. He related this to many situations that I had experienced. When I mentioned that a tall, husky middle aged man with Irish looking features and kind of a beer belly followed me and my friend Irene from a restaurant, Jordan listened seemingly in alarm. He got into a black van with blacked out windows and a near antenna-farm on the roof and followed my car slowly for several blocks, until I said loudly to Irene that if the van followed us any further, I would drive right to the police station. The van driver then beeped the horn and turned into the street on his right. I felt that he had a listening device pointed towards my car and had heard what I said. The horn honking was an acknowledgement of it. Jordan questioned me more about the looks of the man. He said that it sounded like one of his superiors at the Bureau, who specializes in intelligence matters.

Jordan then proceeded to reveal that this superior was the same man who had given the party where his wife met her AFOSI lover. He also remembered that when he was in Atlanta, at the agent's training session, this same boss had phoned to ask him a question. He asked Jordan if he had heard of a particular super secret "black" project. Jordan told him he did and revealed the nature of it, to the best of his knowledge. Jordan thought that it was strange that his boss had called and asked such a question over an unsecured hotel phone line. He asked if I thought that I could recognize the man in the black van, if I were to see him again. I told him that I didn't know for sure, but I would know what he *didn't* look like, for sure. He said that he would try to arrange for me to check out his boss, probably at a restaurant. I could be at an appointed place, when they arrived and look at his supervisor secretly. I asked Jordan when that would happen but he never really got around to doing that. Ever my protector, Jordan was!

During the period of the flurry of telephone message and call interruptions, Jordan asked me if I knew anything about the Jicarilla reservation in northern New Mexico. I told him that most knowledgeable UFO researchers did. That reservation is the home to the most heinous of supposed alien/U.S. military bases, the Dulce base. There was much anecdotal information about a deep, underground base in that area, where Nazi type and high tech genetic engineering was going on. Vats filled with animal, human and hybrid alien/human bodies, alive and in stasis, and body parts of all kinds in test tubes have reportedly been seen. Other weird stuff, like humans sitting in a room at what looked like a computer console doing nothing but repeating sequences of numbers out loud have also been viewed.

There was supposedly a major confrontation between the aliens and the human military special security detachment for the base. Apparently, the human group was appalled at what was going on and an insurrection occurred. Both humans and aliens were killed. It was said that this insurrection was witnessed by the Native Americans on the Jicarilla reservation, including the reservation's sheriff.

Jordan said that he was told to report there as a witness to a murder that had occurred on the reservation. A Jicarilla reservation Indian had murdered someone and the witness to the murder had been living here in South Florida. Jordan had taken the murder witness's testimony here, as he was the local Federal Agent involved. The witness subsequently died and Jordan was the only one who could testify about the murder. In order to do so, he had to travel to the local venue in New Mexico. He said that he would stay at Aztec, a town near the reservation. I told him that Aztec was also famous in ufological lore as the site of a saucer crash. I told him that there was even a book written about it.

Jordan said that he was not aware of those facts. I remarked about the synchronicity of his going to that place. He stated that he barely remembered the case and had to go to his files to cull the case up in his mind. This was going to be his first experience in the Southwestern U.S. and he said that he was looking forward to going. I warned him about going to Dulce, if he, indeed, was being surveilled with me. It would be a good place to "deal" with him, one way or the other. Jordan seemed fearless. He said that he had gotten in touch with the FBI agent involved in Aztec. He said that the agent was very friendly and had a small plane and would fly him anywhere he wanted to go. Jordan thought that as long as he was in the area, he would also wend his way up to Area 51 at Nellis Air Force Base, near Las Vegas, Nevada. This is another area of unusual night time activity of saucer-shaped craft and high technology research and development. Many locals feel that there is reverse engineering occurring there, where retrieved alien craft are being taken apart and flown by U.S. military, in our attempt to understand their technology. *(See PANDA UNCAGED, page 3)*

I told Jordan that Las Vegas, Nevada and Aztec, New Mexico are not exactly neighboring towns, but are hundreds of miles apart. I also asked how he was going to get on a highly secured base such as Nellis, without a need to be there. Jordan told me that he had already made contact with what he called the "Dreamland" police and they had okayed his arrival. He said that the police contingent there was private and not military. I remarked that I already knew that and that it has been said that Wackenhut Corporation, a huge private security firm, was handling the policing of the area.

Synchronicity aside, I found it incredible that Jordan had entree to two of ufology's greatest mystery areas, without any major hassles. I especially found it strange for that to happen so easily, when he was also supposed to be an object of interest and surveillance along with me. Was he trying to find out what I knew about these areas? They were knowledgeable to me only because so many others have written and researched these areas. I certainly had no first hand experiences there. Why was it that, out of nowhere, he was all of a sudden going there? This was another part of the riddle within an enigma.

85

One Wednesday afternoon, Jordan came to take me to lunch and to "enjoy the pleasure of my company," as he so gallantly put it. I also enjoyed being with him. I felt as if there were two parts of my persona, one that loved his soul unconditionally and completely, and the other that listened to everything he said for glaring mistakes, errors, omissions or sinister plots (the sinister plots, of course, being directed at me). I wasn't being paranoid, I was just being careful and alert. He seemed too good to be true. On one hand, he was very helpful and intensely concerned with my well-being, and on the other hand, he was still the FBI Special Agent who loved his job and was good at it. I knew that I was being set up to trust him completely, which I absolutely *did not* feel. I was catching on to this ruse. The fake bug in my smoke detector was an especially glaring error on his part.

There was something in his way of being that made me feel close to him. I often mentioned that we had a bond, no matter what happened. I said that mainly to stay at least neck and neck with him in the game. However, another part of me, despite all of my rationalizations, did believe it strongly and unconditionally. Was it just karmic reality or something else? I kept on pondering whether we both were under some kind of mind control or was it just me. It wasn't because I felt that I was doing, saying or acting out of keeping with my normal mode of being, but because it seemed that I so often knew exactly what he was going to say and how he was going to react. I sensed that this surpassed my own psychic boundaries.

It felt like something else was at work here. I wondered what it was. There were also times when Jordan stated that he knew what I was going to say. One day, we were discussing interdimensionality. I started to relate the paranormal experience I had in Kenya when I was on a safari. Jordan immediately got pale and started to hyperventilate. I thought he was going to have a heart attack. I asked him what was wrong. He said, almost inaudibly, that he knew exactly what I was going to say and swore that I had told him this story before. He went on to say that he had *never* before had such an intense feeling of deja vu, and it scared him so he could hardly speak! If it wasn't for his real physical reaction, I would have thought that he was just putting me on. Our relationship was short enough that I knew I had not ever discussed this before with him. Such was the tenor of our friendship. It was odd but powerful, ironic but seemed filled with heartfelt meaning. Only God knows what we truly are to each other.

Also, the way he communicated verbally was so similar to my patterns and use of language. I often felt like he was my alter ego. I was so comfortable in his company. Jordan said many times that he felt so "at home" with me that he had to really force himself to leave my apartment. He seemed so childlike, like a kid who didn't want to leave Disney World. That reluctance to leave seemed genuine. I never encouraged him to stay for lengthy periods, since I didn't feel comfortable when I left the room or he did, even to go to the bathroom! I thought he might do something in his G-Man posture covertly. Somehow, I felt comfortably schizophrenic about this. I did not want to opt out of this chess match. I also knew that even if I wanted to opt out, I doubted if "they" would let me. I didn't understand why I wasn't really frightened. Sometimes, I felt eager, like a Special Forces Blue Beret team member. There was something within me that had prepared me for this "battle" and I was energized for the fray! During these months, I had felt very alive, all of my adrenaline seemed to be pumping full time. Yet, I had an inner calm and clarity. I knew that somehow, this had to happen. Within this puzzle within a puzzle, I was hoping to find my answer about the intensive years' long surveillance.

A mental and problematical turning point occurred during one long afternoon's visit. Jordan said that he had taken a vacation day off, just for this time together. He was dressed, neatly and fashionably, as always. He also carried his weapon in his black fanny pack, as he always did. He asked permission to take it off, and laid it on my cocktail table. It was very heavy, cumbersome and uncomfortable. I also felt uncomfortable seeing it, but again I felt that strange duality of uneasiness and excitement. Was Jordan really a vulnerable super cop or something else? The gun was symbolic to me of this very big question. I didn't think that his intention was to hurt me, since he had plenty of opportunities to do that, with or without his gun. Almost every time he saw me or spoke to me on the phone, he always spoke affectionately. Initially, his words were that he was falling in love with me, later on it was that he did love me very much. Not that I doubt my own lovability, but his impetuousness in falling in love seemed too rapid. It did not seem appropriate for the scenario, but he was a very apt Lothario. As much as I felt this unique bond, I would never allow myself to fall in love with a question mark, let alone a married, albeit unhappy, man with young children. His constancy was not contingent upon my return of committed affection. I was more concerned about his real and true intentions and reasons for romancing and befriending me.

Prior to that afternoon's visit, I had, what I call, a "beeper.". It is a powerful intuition that something is going to happen. In my mind's ear, I could hear Jordan remark about a large, framed black and white photograph that hangs in my apartment. I have had this photograph for 14 years and no one has ever remarked about it. It has a surreal feeling to it. There is a large abstract form and hanging on to it is part of a female nude body, side view. There is no head seen to identify the woman. It is very arty looking, more illusion than substance.

I *knew* that when Jordan would come into the room, he would remark about it. I could hear him asking me if that was *me* in the photograph. He had been in this room before with the photograph but had never commented on it. This time, just as I had envisioned, as soon as he walked into the room, he pointed to the photograph on the wall and asked if that was me in the picture! I just smiled and said no, but I was a bit shaken. I had not discussed this with anyone, yet he said exactly what I knew he would. Just like so many times before, we were saying and doing things the other already knew about! There was a high level of communion or connection between us. What could it be? I just couldn't stop thinking that perhaps it *was* some kind of mind control, or that we had done all of this stuff before, somewhere or sometime else. Was it in this plane of consciousness or another? I was baffled but felt challenged to keep the game going, that is, if "they" would allow it.

Our afternoon's visit passed quickly and, as always, enjoyably. Jordan was never at a loss for conversation. He discussed his cases at work, his children and his personal problems, including what he was now calling, *his* surveillance by unknown entities. He also was concerned about a regressive hypnosis session he was going to have in 3 days, to try to find out about his post traumatic stress syndrome. Jordan said that his counselor was quite insistent that he try it to get to the crux of his all-encompassing anxiety. He repeated over and over that he was sure that it was because of his wife's betrayal of him with the AFOSI guy and not for any other reason.

He mentioned the name of the therapist who was going to do the hypnosis and asked if I had heard of him. I told him, I had not. He said that he wanted to get together with me right after his session to discuss the results of the regression. I agreed to be available.

It was on that day that our relationship escalated. I had decided that after all of the intellectual and emotional considerations of the prior months, perhaps it would be best to move the relationship to a physical level. The decision to do or not to do took up too much time and energy. I hadn't put so much pondering into this kind of decision since I was a virgin! I thought for sure Jordan would be happy about this. He had been talking and fantasizing about this liaison. When I announced my decision, after all the months of saying no, he was not as enthusiastic as I had expected. All of a sudden, he had a paranoid feeling that we were going to be taped by "them" and his children would see us via their VCR! Despite that, he participated in a rather joyless and apologetic (on his part) mating. I was not surprised at his performance anxiety and reassured him that it wasn't important. Indeed, it wasn't. I was just glad to get past that barrier. Something instinctively told me that it had to be done for my own good.

Afterwards, Jordan commented on the fact that his beeper had not gone off even once during the afternoon. I commented that my phone had not rung in several hours, either. We were getting ready to leave the apartment, as I had a meeting to attend and Jordan had to coach his son's soccer match. Then his beeper went off. Politely, he asked permission to use the phone. I consented and went to get my keys in preparation to leave. I only vaguely heard his conversation, since I was otherwise occupied and tried hard not to snoop. I often feared that I might hear something top secret that I was not supposed to and that something would happen to me if I did. That tactic had been tried on me before, with the mastermind, master criminal, Corey. Thank God, I had the insight to catch on. Therefore, I had no interest in listening in to Jordan's conversation.

In a few minutes he finished and as we were walking out the door, I asked him if he wanted to come to the garage to inspect my car for listening devices or tailing "bugs." A few weeks before, Jordan had suggested on his own that he do that. He had said that he might do it without telling me but that I would find something from him to let me know that he, indeed, *had* been there. After I made the request, he looked at me in surprise. "Don't you know that I already did that? Didn't you see my business card on the front seat of your car?" "No," I said, "when did you do it?"

"I came by last night at 11:30." he reported. "I had to wait in the bushes for a long time before somebody came into the garage. Finally, someone did, and when I went to get into your car, with my master car key, I had a hard time but I finally did it. Then when I finished checking the car out, I couldn't figure how to get out the garage door. After awhile, another car came in and I snuck out. I checked out your car thoroughly and you don't have any bugs in it."

"I haven't been down to my car today," I replied, so I guess that's

90

why I didn't know that, although I am surprised that you didn't see the garage door opener on my car's visor. Also next to the garage door are big buttons to open and close the door from inside." "I didn't notice them," he said casually. I definitely was not feeling casual about this conversation. He was patently lying!

We said goodbye in the lobby of the condominium. Jordan was parked in front of the building and I went down to the garage. I was getting ticked that Jordan was lying again. Firstly, he had mentioned early on in the afternoon's conversation that he had tried to call me at 11:30 the night before, but I did not answer the phone and my answering machine did not take the call. I told him that I was at home and that the phone did not ring at that hour. He didn't mention anything about being down in my garage then. He must have forgotten he mentioned that to me, just four hours earlier. Why would he travel two hours round trip to check out my car, when he knew that he would be there the next afternoon and could do it then?

Although the garage door is opened by a garage door opener, there always is a lot of traffic in and out of the garage and he would not have had to wait a long time, even at that hour. There is also another door adjacent to the one I enter, and he could have easily gotten to my car through either door. I also found it hard to believe that a certified Federal Agent wouldn't be savvy enough to notice the garage door opener in my car or the buttons to get in and out of the garage. These things were not hidden! I also checked my car thoroughly and could not find his business card anywhere on, under or between the seats. Why would he lie about something like this? There was no obvious reason for it.. Was he losing it? Just what was the point? Surely, he should have known by that time that I would look for his business card and know that he had *not* been there.

I then had an awful feeling. Another one of those "beepers" went off in my head. Something told me to go up and check my tape recorder that records all of my phone calls, incoming and outgoing.

I truly had not thought to check up on him in this manner, and as I mentioned before, I really was afraid to listen in on his phone calls. But now, I knew that I had to do it!

My instincts about that were correct. When I played back his conversation, I felt sick. I couldn't believe my ears. After entering a three-digit code, he entered a full phone number... and that number was busy, at least it had a busy signal tone! As I listened, I heard Jordan speak for several minutes to a busy signal. His conversation was to indicate that he was speaking to his boss about his work. He was very good at improvisation. He would pause, as if he was listening to someone on the other end of the line, and would make comments and statements as might be appropriate for that kind of conversation. He carried this charade on for over three minutes before he hung up. I remembered that as we were on the way out, he had mentioned that his boss never beeps him when he is on duty but always does when he is not!

As I was listening to the busy signal conversation, I wondered whether this could be some kind of device that the FBI uses to keep from being surveilled over the phone. That would not seem to apply for this scenario, however, since the tape recorder recorded the conversation from the same phone from which Jordan was speaking!

Why would he talk to a bogus boss in a bogus conversation? Jordan didn't need a reason to leave. I wanted him to go, and he had planned to, anyway. After the conversation, he didn't say that he had to get somewhere fast. The only possible motivation I could think of was that the initial three digit code might have triggered a signal for something or someone else. This was really crazy! Just like the stupid garage story lie, this phone conversation lie also seemed to make no good sense. Then again, maybe you have to be a crafty Federal Agent to comprehend the rationality of such things.

I was beginning to think that Jordan was a loose cannon with a lot of rattling nuts and bolts. He made condomaniac Mark look sane!

For a guy who spoke so calmly and so articulately, with a perfect demeanor at all times, he was making some grievous errors. Even though he knew that I tape recorded all of my phone calls, as a precautionary method of protection, he must have forgotten this very important fact, when he made his bogus call to the busy signal. I wondered what he would do when he *did* remember!

I phoned my friend and confidante, Susan, and played Jordan's crazy busy signal interchange to her. She said that she would come to my apartment to discuss this, since we didn't want to talk about it over the phone. At 8:30 P.M., we were deep into the discussion, when the phone rang. Serendipitously, it was Jordan. His calling at this hour was quite out of his normal mode. I knew intuitively that he had finally remembered that I tape recorded phone conversations and was calling, on a fishing expedition to see whether or not I had played back his call. Even though he had just left me four hours before, he said he was calling just to see how I was doing and to thank me for a great afternoon.

Oh, sure he was! He then went on to tell me another equally creative tale about what happened after he left me in the condo lobby. Jordan said he got in his car and drove to the first stop light, before he noticed a plain, white sheet of paper, tucked under his windshield wiper. He pulled over to a parking place and retrieved the paper. It was a plain, white sheet with a series of six single-digit numbers. In between each number was a dot. It looked like some kind of code. Jordan said that he returned to the condo's parking area to see if any of the other cars had a piece of paper under their windshield wipers. None did. So, he assumed that the code was meant for him. He said that he was going to bring it to a friend who was a retired FBI cryptographer for analysis. Jordan said that he looked for me in the condo lobby, but I wasn't there. In this instance, he was right, as I was down in the garage looking for his non-existent business card in my car!

So, here again was more intrigue, more sly machinations. The wonder of it all was that I was the lucky beneficiary! I decided that

I would keep my cool and not yet divulge to Jordan my long and growing list of grievances about his credulity. I knew the right time would present itself for me to show my hand. I wanted to catch him off guard, unprepared with rebuttal. I had a hunch that he was prepared at that moment for my condemnation because he kept asking me if I was alright!. We passed some chit chat and the conversation ended with Jordan saying that he, was looking forward to being with me again in three days.

Jordan called again early the next day and left a message on my answering machine. Again, as in times before, the message had been edited by unseen forces. On my tape recorder, it was full and complete. When he phoned later in the evening, our conversation was accompanied by a variety of beeps, tones, and clicks. One time, our conversation was cut off in mid stream and we were disconnected. Prior to the disconnection was a long tone. We were deeply engrossed in a debate on Jordan's regression scheduled for the next day, when we heard a loud metallic clatter. Jordan said that it came from his backyard.

We were obviously not going to be allowed uninterrupted repartee. Jordan got angry with all of these interruptions. He said that those who were watching him were "making a lot of mistakes" and he was "on to them". I asked, "Well, how does the watcher like being watched?" He said that he didn't, and felt he was getting paranoid.

He stated that he was totally committed to finding the truth, or at least a piece of the truth about UFOs and extraterrestrials. He said that he wanted to find something that could be verified, some assurance that they are real. When I heard what he said, I laughed. "Lots of luck, bro. How do you think you are going to do that when they can manipulate our phone calls and keep us from talking to each other? To get to the larger truths, you need a lot more than good intentions. You need to be in the seat of power. I doubt if we can do anything more than "they" will allow."

Jordan replied, "I am determined and will hang tough doing it. 'They' are trying to push me in the other direction but it only propels me more towards finding the truth." His righteous words were ironic to me. He uttered them with great feeling. I asked him if his bosses were giving him a hard time because of his new focus. He said that they weren't.

"Did they mention anything about your being surveilled by other intelligence agencies?" I asked. He claimed that no one had said anything. "Don't you think that they would have to be involved with the spying on you, or brought into the picture somehow?" Jordan sighed and said that he didn't know but couldn't see any changes in their attitudes towards him, although he did have an unusual conversation with one of his bosses. Since our relationship had begun, Jordan said that he had continually used the FBI data base to access information of any and all kinds related to the UFO phenomena. Almost each work day, new information or closed files found its way to his desk. One data byte that really intrigued him was reference to Division 5 within the FBI. Apparently, it was a secret division, even from special agents. Jordan asked his immediate supervisor if he ever heard of Division 5. His boss looked at him with surprise and asked how he knew about it. Jordan said that he was just referencing material related to his "X-File" cases and the name of that unit came up. His boss briefly related the focus of Division 5 as being super secret and covert intelligence gathering of a highly specialized nature. The next day, the same supervisor told Jordan to cease and desist from accessing information related to extraterrestrial phenomena, since it was a waste of his time and he should address his normal caseload.

Jordan's conversation again returned to his concern about doing the hypnotic regression, the next day. Although Dr. Weeks, the psychotherapist who would be doing it, was highly recommended, Jordan was afraid about what this might stir up. He said that he was glad that he would see me right afterwards because he felt comfortable discussing the results with me, whatever they might be.

His statement presented the perfect catalyst for me to begin to say what I had been wanting to say. "Yes, I am looking forward to the results, too. I also have some very important things that I want to discuss with you that have been bothering me. I need you to clarify some stuff that is unsettling for me. You must be absolutely honest with me, even though I know you have a job to do."

Jordan responded quickly, "of course, I will be truthful. You have my total loyalty. My feelings for you supercede my job, you are that important, but what is it that you want to tell me?"

I responded, "I don't want to discuss it over the phone. You have a big day tomorrow with Dr. Weeks. I don't want to stress you out any more than you already are. It can wait until tomorrow. I think that we should not meet here at my apartment. If you have any super personal information, I think we should talk elsewhere, since we know my apartment most likely is bugged."

Jordan agreed that would be best and again inquired as to what it was that I wanted to discuss with him. He asked if I had changed my feelings about him, in any way. I again told him that all discussion would be best put off until the next day, to insure not complicating his regression. Jordan was not happy about that. I felt that he sensed I was going to talk about all of his mistakes that I had picked up on.

Meeting in a place other than my apartment was for my own safety. I was afraid if I confronted him in person and looked him in the eye, he might panic. I didn't know what he would do about that. Even though he always was very gentle and courteous to me, that might be a facade, an act. There also was the chance that he was under some kind of mind control. If he were challenged, perhaps he might act in a programmed, predetermined way that was aggressive. He did have a gun and was most likely a total spy. For my own security, a place with a lot of people would be more apropos for a showdown. At any rate, I was greatly anticipating finding out the truth, or at least a part of the truth, about Jordan's activities. I also was very nervous about the potential encounter; and Jordan's regression results were the least of it.

I rose early the next morning and made time to meditate. I wanted to be centered, cool and resolute, when I met with Jordan. I knew he had a 9 AM appointment. He said that it should take an hour. When it was past ten, I began to get fidgety. "He better call," I thought, "I can't take the suspense!" At 10:30, the phone rang. I said hello twice and then heard a male voice say, "Oh, shit," after which he hung up. I couldn't determine whether that was Jordan or not, it was so fast. Jordan rarely cursed. If it was him, why would he hang up? I waited another half hour and then decided to beep him. Soon after, the phone rang and it was a barely audible Jordan who was on the other end.

Jordan was calling from a phone booth, still in Miami, an hour after he finished his appointment. I asked him if he had phoned and hung up. Jordan confirmed that he did. He said that he was too freaked out to talk and had to get a hold of himself, before he could. I asked him if there was anything I could do to help him. He responded with a diatribe against some unknown force or forces that accounted for his current misery. I was also blamed in his monologue.

He began, "Dr. Weeks thinks I am crazy. He thinks I made up it all up. I know it was real. I was six years old when it happened in New York. I feel like I knew this was coming. I am being used and I don't like it. They took me and they weren't the little guys either. They were as big as grown ups! I wasn't suppose to remember anything and I don't want to remember anything else! What are these things? They talked to me and then they took me! Oh my God, those big eyes! I am so scared! I don't want to do what they want me to do. I don't want to be moved around. I don't feel safe at all. They were real and they *touched* me!"

I tried to calm him down, asking him questions, but he wouldn't have any of my questions. He sighed, shuddered and cried.

I asked him who the "they" were that want him to do things and have him moved around, were they aliens or FBI? He said he didn't know. I said, "Knowledge of your experience can empower you. It is much better than ignorance. At least, you now know the source of your stress syndrome and can learn how to deal with it." Jordan screamed, "I don't *want* to deal with it! I can't take it like you take it. You're very relaxed."

I replied, "I am not relaxed, I am logical and have learned how to cope. It is not aliens that scare me, some humans are a lot more ferocious. I have not been intimidated openly or been followed by aliens. It is humans who I have learned to deal with and who have violated my constitutional rights of privacy over and over. The worse thing for me is that it doesn't seem that I can do a damn thing about it, except to learn how to live with it and remain sane and happy."

The rational approach fell on deaf ears. Jordan continued his hysterics unabated. "I don't know who to trust. I didn't want, ask or volunteer for this! They took me and I want them to stay away from me. I want it to be true that they only came that one time. I want to have a life, everything is wrong. I don't feel safe, I am not going to let them .... I'll just ... I don't want to be helped with this shit! I won't let them get away with this!"

I tried to calm him down again. When I said that I will always be there for him, if he needed someone to talk to, he turned on me. "I don't want to go anywhere *near* you," Jordan declared, "this is connected with you, this is you!"

"What are you saying," I replied strongly, "that I'm an alien?" Without answering me, he went off on another tangent. "If you're listening, you should have told me!" he said in a choked voice. I assumed he meant the agents that we both presumed were listening to my phone calls on a continuum. Jordan's departing words were, "Screw them and screw this whole thing!" and then he hung up. He didn't even say goodbye to me.

Stunned, I looked at the phone. I thought, "Is this the way he and or 'they' are going to end this relationship?" It would be a pretty facile way of doing things. If Jordan had a hunch that I was going to quiz him about his faux pas' and he didn't have good explanations, this could be his way of getting off the hook. His histrionics were totally *not* in keeping with his normal cool demeanor. This was a performance I had never envisioned for him. I really was not convinced of his sincerity in the pain he expressed. He had read a lot about alien abductions and could have easily used the information for his own abduction episode.

In a way, I was relieved. I wouldn't have to face him now with my own grievances. His fast "kiss-off" of me, blaming *me* for his problem, assured me that I would probably never hear from him again.

Later, I checked with Dr. Weeks, Jordan's supposed psychotherapist/ hypnotherapist. I pretended that I was a potential patient, calling for information. I asked Dr. Weeks if he did hypnotic regressions, as part of his practice. Dr. Weeks was emphatic in his response. He said, "Absolutely not! I would never do that. It is worthless. If you want to really make progress, you come to me. I use sound therapeutic techniques, not hypnosis. That is for charlatans!"

Dr. Weeks' response was interesting. With his pretentious, closed-minded attitude, he certainly was not the kind of practitioner I would choose. And if he was telling me the truth, it appeared that Jordan didn't use his services either. This was another one of Jordan's strangely blatant mistakes. Why would he tell me the name of a real therapist if he wasn't going to use him? He could have made up a name and I would have been no wiser Again, did he underestimate my intelligence and think I wouldn't check up on him? It was then safe to assume that yet another charade had been offered to me on this smorgasbord of charades, and this one was equally unpalatable! The phrase "intelligence agents" is truly an oxymoron!

*"Mia, I have to be very careful.
I don't have very long to talk.
I don't know what's been
happening to you, but a lot's
been happening to me.
You have got to try...try to
remember what happened to you
in New York around 1962..."*

– Jordan Perez
FBI Special Agent

POST REGRESSION ODYSSEY ODDITIES

I really didn't think that I would ever hear from Jordan again. He had certainly made his kiss-off sound final. All of my friends were sure that he would once again return. That remained to be seen, I thought. One of my personality quirks, and not my best trait, is my need to have the last word. Since Jordan had hung up on me, I felt it necessary to deliver my undelivered communication, unequivocally. I wrote:

Sweet Jordie:

Despite your incredible duplicity and arrogance, I do NOT hate you. In fact, I found the last three months to be very exciting and memorable! There were parts that were downright enjoyable. I am smiling to myself as I write this. This must be the perverse part of my nature. From the get-go, you must have thought that you were really putting something over on me. I am sorry to tell you, I have always had doubts about your focus. Like you, I notice things. There were a lot of chinks in your armor. It is too bad that you are such a scoundrel, as I really enjoyed your company. It is interesting that you excoriate Andy, since you are a different side of the same coin.

Being the object of your surveillance, was always in the equation of possibility. I forgive you and the others who glory in their sadism. I don't care if you don't want my forgiveness. I am not your willing victim in some idiotic G-Man game. You made a lot of mistakes and now are checkmated! Will you continue the game or are you like the nutty ex-FBI guy you supposedly investigated?

The thing that bugs me is how did I know you so well? Was it just my innate instinct and intuition or is it something else? I did have a glimmer of hope that you were the good and decent person you tried to appear to be. There was an instant when I thought it might be true! But, alas, it was only an instant!

No matter what the truth is, on a soul level, my soul sends love to your soul in recognition of what has always been there. Of that, I have no doubt or equation.

Much more than sincere,
   "THE BRIDE OF THE COSMOS"

P.S.
I guess I'll never find out why you guys think I am so damn important!!!!!!!!!!!!!!!!!!!!!!!!!!!!

After I wrote the letter, I was sure that it would dot the i's and cross the t's of our relationship. I phoned his office to check to see if he was still there. I had this feeling that he had been withdrawn somewhere into the bowels of intelligence agents' purgatory because of his mistakes. When I was put through to his voice mail, the only voice that I heard, was an automated voice. Jordan's own voice giving his name and title was missing! I found that odd, since it deviated from the norm. I did not hear from Jordan for a week and a half and then he phoned me. I was very surprised, especially since I had excoriated him so severely in the letter. He didn't mention the letter and I was too shocked to remember to ask if he had received it. The tenor of the call was very "cloak and daggeresque." This is how the conversation ensued.

(The following are transcripts of the actual tape recorded conversations between Mia and Jordan with Jordan's full knowledge that all conversations were being recorded.)
MIA: Well, well Jordie.
JORDAN: *(Speaking rapidly in a hushed, stressed tone)*
               Mia, I have to be very careful. I don't have very long to talk.
               I don't know what's been happening to you, but a lot's been
               happening to me. You have got to try...*try* to remember
               what happened to you in New York around 1962.

102

MIA: New York in 1962?

JORDAN: You have *got* to... you have *got* to focus on that.

MIA: I wasn't even *in* New York in 1962.

JORDAN: Alright, well...I don't know *where* you were, but that has
           got to be something you have got to figure out. I don't
           have long to talk, and (sigh) are ... are you okay?

MIA: Well, I'm not really happy about what you've been
           doing!

JORDAN: What I've...(exasperated sigh) Okay, just *please* think
           about that and I will be in touch... I will definitely be in
           touch...You will know that it's me when the time comes.
           PLEASE! PLEASE! PLEASE think about 1961-1962.

He hung up, without a goodbye

I was flabbergasted! There was so much I wanted to ask him but he
was like a wordy, whirling dervish. He made no reference to the
regression debacle, my letter or anything other than my having to
remember 1961-1962! Since, I was in college at that time and knew
exactly where I was, remembering was no big deal! During that time,
I was in college in Tampa, Florida, at the University of Tampa. It was
the one year that I could afford to go away to school. My parents
had bought what then was called a "penny policy." It was a simple
term insurance policy. I think I got $500 or $600, on my 17th
birthday.. In those days, colleges were affordable, even for lower
middle class kids like me. I chose Tampa U. for a few simple,
uncomplicated reasons. Primary was the weather. I detested waiting
for buses and elevated trains in bone chilling cold Chicago weather.
Secondary was the student body cross section from all over the
United States, especially the Eastern U.S. I loved those New York,
New Jersey and Pennsylvania guys and how they danced. I had never
been to Florida before, so I chose it for the warm weather and the
hot dancers!

The vestigial remains, before and after time-wise, were spent
assuredly uneventfully, at home in Chicago, doing what I had done

103

for all the boring years in the past. What on God's green earth was the man talking about, that could be connected to these years' long surveillance and intrigue?

Despite this enigmatic statement, I did have a thought of what he could be talking about. It was an immediate response to the stimulus of the date he gave. I had a very strong image of an experience I remembered, even after many years. I never forgot the emotion of the experience. Sheryl was my friend who lived in my dorm and had a shiny Seafoam Blue Chevy convertible. It was gorgeous. We loved to bomb around in the car, visiting all of the local college hangouts. One evening Sheryl said that she wanted me to come with her to a local Tampa beach. Her intention was to meet up with Randy, a fellow student she had a crush on. I protested that it was getting too dark to go there but she insisted that it would be for just a few minutes. In my gut, I did not have a good feeling about this but I relented. As we approached the isolated beach area, there were no other people there except Randy and his shiny white Corvette. I was really feeling uneasy as we approached Randy.

When we got out of the car, I saw that Randy was looking in the trunk of his Corvette for tools. I asked him if there was something wrong with his car, he replied that his car was fine but he wanted his tools and car jack for something else. Randy took the jack and his tools and proceeded to walk over to a boat trailer that had been left on the beach. He said that he was going to steal the tires from this boat trailer for his car, since they were the same size as his tires. When I heard that I really began to panic. I begged Sheryl to leave, since I was sure we would get caught by the police and that I would get expelled! I don't remember how soon we left after that but I do remember the intense fear, almost like terror, I had. I have never forgotten that incident, even though it was more than 30 years later!

Could it be that this experience that has left such an indelible mark on my psyche was really a cover memory? Could this really have been some kind of abduction experience? Perhaps since the beach was close to MacDill Air Force Base, if there was an abduction, it could have been witnessed by radar or humans. During my year in Tampa, I did spend time on the base, since I had a boy friend Bobby Robertson who served there and I went to dances on the base. Was this the link in the chain of events in my life that included Jordie?

I had two days to think about that before Jordan called again.

JORDAN: Mia, how are you?

MIA: How *am I?* (laughing) Don't you know? How am I? I'm fine.. I'm good, the question is how are *you?*

JORDAN: I don't know...a lot's been happening, but a... I just wanted to make sure you're okay.

MIA: Yeah. You really care if I'm okay or not, huh?

JORDAN: (pause, big sigh) Yes.

MIA: You do, huh?

JORDAN: Yes.

MIA: Well, maybe you'll explain some of the strange things that you did.

JORDAN: (sigh)

MIA: Are you telling me that you're *not* a spy...that you're not spying on me?

JORDAN: Mia ... I don't know what you believe or not believe, and I don't know what's been happening ... (breathes another sigh) but I just wanted to tell you that ... uh that the day will come that you *know* everything and that you...have absolute certainty that everything I've done and everything that I've said to you ...was true!

MIA: Well, I sure hope so!

JORDAN: ...and that ... that, um ... you can see it.

MIA: You think that will happen, huh?

JORDAN: Yes

MIA: Are you convinced that there are others out there that seem to have an interest in me, for some strange reason?

JORDAN: Yes.

MIA: Do you know *why* they do?

105

JORDAN: What I told you the other night ... do not ... (pause) what I called you...don't ... let's discuss that right now.

MIA: Okay.

JORDAN: But that...*that* is the center of everything!

MIA: It is...So, I would imagine if you know that, there must be some record of that.

JORDAN: (pause, sigh,) Quite a bit.

MIA: Hmmmm. Well, apparently, they've done a good job of helping hide it in my mind, because I really have no recollection of anything extraordinary.

JORDAN: That's the problem.

MIA: Why are you so cryptic

JORDAN: Because, I believe that right now is a good time to call you, but I'm not sure ... and all I ask is that you be patient and ... keep your eyes open... that's all I can tell you.

MIA: Are you saying that I am in jeopardy?

JORDAN: Not to my knowledge.

MIA: Okay, then you mean something else, that I will see something that will be ....

JORDAN: From me, yes, from me.

MIA: I just would hate to think that everything that you ever said was not true!

JORDAN: Everything I said...what I said, at the time I believed to be true ...and for the most part, as far as I know, it was true.

MIA: What about your own feelings? Were they orchestrated?

JORDAN: (pause, in a very emotional voice) Mia, I *love* you. I don't know how I could be clearer!

MIA: Well, that perplexes me in that there were a few things that you did that didn't really show that.

JORDAN: Things that *I did*?

MIA: Yeah, that I know were not true. I don't know whether there are certain people who think that I am totally naive or have underestimated my intelligence, or whether this is just some big game and I'm a rat in a maze and they're watching my responses.

JORDAN: I think nothing *could* be more true than that...what you
just said.

MIA: It was never created to be subtle. It was always arranged
to be overt.

JORDAN: I didn't see it until...(sigh)...'til it hit me like a brick.
(Pause) We're not alone in this, and ... I'll do what I have
to do....

MIA: And you think that I will know sometime what this is all
about?

JORDAN: I'll do everything in my power to make that happen!

MIA: Is there any connection between you and me in this or is
this just happenstance?

JORDAN: (ironic chuckle) There couldn't *be* a stronger
connection!

MIA: Then you know what the connection is ...?

JORDAN: Yes... and this has to do with what I mentioned the other
day.

MIA: Yeah, but you were just...were you born... were you born
then?

*(Jordan breathed another audible, emotional sigh and quickly
hung up the phone.)*

I have heard the phrase "being thunderstruck" but it never had any
real validity in my life before. The moment Jordan hung up the
phone, the proverbial light bulb went on over my head and I felt
thunderstruck! "Oh, my God," I said to myself, "it couldn't be
possible! But it would explain a lot of things and leave a lot of others
unanswered!" I sat there shaking my head no, over and over. The
only clues Jordan gave me to the riddle in my life is the year 1962 or
the years 1961-1962, and his stating that we couldn't have a closer
connection! Jordan told me that he was 32 years old, when I first
met him. That would mean that he was born in 1962. Could he
possibly be alluding to the fact that *I am his mother!* If we have a
cosmic connection, could it be through alien ovum harvesting?
Could he be the result of my egg and some kind of alien genetic
incubation? It is much too bizarre to contemplate. But why else

107

would he have panicked, when I mentioned his birth date?

Or perhaps, "they" wanted me to think that and wanted to see how I would react to this fact. It would be especially interesting, considering our budding romantic entanglement. I certainly did not feel maternal towards Jordan, although I did feel a close, unexplainable bond. When I was married, I was not able to have children. I went through months of infertility testing and medical rituals to get pregnant. My doctor was at a loss to explain my infertility. There did not appear to be any known causation for the problem. I was one of his few patient failures. So, if this was true about Jordan, it would be a very ironic twist to my life. I felt no guilt about my feelings towards Jordan. I wondered if Jordan felt guilty about any potential Oedipal conflicts. The ball was now in Jordan's court to explain this weirdest of weird scenarios.

A few days after Jordan's "birthday" call, I received a very unusual message on my answering machine. A tinny sounding female voice uttered six times over an address that I had never heard of. At first, I thought it sounded like a clear dispatcher's voice saying "12400 SW 177 Avenue". I could hear the tape stop and begin again, playing faster and faster each time. After the sixth time, it was very difficult to hear the address distinctly, unless you knew what to listen for. I presumed the address was in Miami, since Ft. Lauderdale didn't have that address. The site of that address was a remote area and the land had been developed as a truck farm with existing crops. It was near a detention camp for Cubans and Haitians, and not far from Homestead Air Force Base, but there did not seem to be any relevance to me. Perhaps, this was another red herring. I know that "they" did this recording but why? Could the woman who recorded this be the same one who recorded the MUFON hot line phone call, a few weeks before? More inanity from the agencies for insanity.

That evening, after the unknown address recording was left on my answering machine, Jordan called once again.

JORDAN: It's me. How are you? Don't discuss substantive issues, just tell me how you are doing.

MIA: I am still filled with a lot of questions and today I got a...

JORDAN: Let's not discuss that just yet...I'll be going out west for awhile. If I give you a phone number and I don't get back will you be kind enough to call that number for me?

MIA: Okay, but how will I know it if you don't get back?

JORDAN: You'll know. I just want to make sure if it is alright with you. You will get everything that you will need before you need it.

MIA: I think I know what you were talking about, the last time we talked and it is confusing.

JORDAN: Let's leave that stuff out. Just keep in mind my promise to you and that I will carry it out.

MIA: Is this going to be our last communication?

JORDAN: I certainly hope not?

MIA: Aren't you going a little early to that place out west?

JORDAN: I know at this point in all likelihood, you have written off most, if not all, I have ever said to you. But this is to be expected.

MIA: It has nothing to do with the way I feel about you, but for substantive things that have occurred.

JORDAN: For whatever reasons, I must operate under that conclusion and rightly so. I know that there are certain things that have been communicated to you and I don't know what version has been given to you ....

MIA: Wait a minute! Nobody has come to talk to me...

JORDAN: No one has said anything to you? No one has communicated to you? *(sounding incredulous)*

MIA: Absolutely not! Why do you think that? I wish someone had!

JORDAN: Okay .... Wow!...Well, then I don't know what to think. You will have an opportunity to reevaluate a lot of things, hopefully for the better!

MIA: I wish that I could talk to you about a lot of things.

109

JORDAN: I have not been available since last we were together. I haven't been around and I want to keep it that way. That is all I have to say. I just hope that you will keep your eyes and your mind open as best you can.

MIA: My mind, of course. I have a lot of gut feelings that can be explained from your last conversation... but if what you inferred is really true, we have to talk!

JORDAN: I don't know what is true quite yet and I am trying to determine it, but I also want to give the people who are listening to us the hardest possible time that I can.

MIA: They're still playing the game.

JORDAN: Yes, I know that and I have been making it hard for them. I have been on the phone much too long now anyway. I am fine but keep your eyes ears, mind and heart open for a little while longer. Be patient for a little bit and know that I love you!

Jordan's referral to going out west, I felt, referred to his trip to Dulce, New Mexico and eventually Nellis Air Force Base near Las Vegas, Nevada. Why would he still be going there, if he was running from some negative "forces?" A few days after his call, something told me to phone him at the FBI office. This time his voice mail did have his own outgoing message. It said that he would not be in for some time and if there were any messages, one should call...and then he gave a number in the 702 area code. I looked in my phone directory and it showed that 702 is the area code for Nevada. I decided to call the number he was supposedly at, and the person who answered said "security police." I asked him if this number was Nellis Air Force Base and he said it was. I thanked him and hung up. Later on, I had my friends Susan and Harvey each call the same number asking to talk to Jordie. Both times it was acknowledged that he was there and they were asked what their names were.

I guessed that it must be the "Dreamland" police , a civilian security force, since they did not answer "military" or "air police" as would be appropriate. Eventually, I found out my assumption was correct.

(When Susan called, she asked, "Is this Dreamland?" and the response was "Yes.") The main question was why was he going to Nellis *before* he went to New Mexico. New Mexico was his main and original destination, while Nellis was supposed to be just a fun and interesting afterthought. Now, I was really wondering just *who* Jordan was and *why* was he there?

Again another couple of days ensued and I, once again, heard from my old phone buddy. This time, he didn't sound very lively. He started to sound kind of depressed or morose.

MIA: It's you at Nellis Air Force Base.

JORDAN: Why do you say that?

MIA: Your outgoing message on your voice mail said that you were at 702 area code. I knew that was Nevada and phoned the number and lo and behold, it was the security police at Dreamland, Nellis Air Force Base. Why are you there?

JORDAN: You know.

MIA: No, I don't, other than you're a spy. I wish I *did* know why.

JORDAN: It just looks like I'm a spy.

MIA: Yeah, no ordinary FBI guy gets to stay at Dreamland. So, I assume you are no ordinary FBI agent. Are you?

JORDAN: I am ordinary as opposed to what you have in mind. How are you?

MIA: I have my first cold in five years. I'm sure that's because my system's down because of my confusion about you. I just wish you would clear things up.

JORDAN: I'll certainly try to do so.

MIA: When?

JORDAN: It depends on how things go.

MIA: How long are you going to be there?

JORDAN: Supposedly through Sunday or Tuesday, it all depends. *(this was Friday)*

MIA: I don't know if I should be afraid to see you. You might try to hurt me.

111

JORDAN: Hurt you?

MIA: I'm so confused about you and what you are and what you're doing. In retrospect, my close feelings of being bonded to you and some of the crazy things you actually did are at odds with each other and are confusing to me.

JORDAN: What you know and what you think you know are two different things.

MIA: That is what I am trying to figure out. I really want to know what this is about and if we have a connection or not.

JORDAN: I don't want to have to hang up, so let's not talk about that. I will attempt to resolve that for both of us.

MIA: It means everything for me to know. It wasn't just happenstance that you were assigned to this case about crazy Mark, my ex-FBI neighbor.

JORDAN: No, there are things I can't talk about now.

MIA: Are you at Nellis because you were made to go there?

JORDAN: No, that was my doing.

MIA: You asked to go there?

JORDAN: Yes.

MIA: What can I know about?

JORDAN: You can know that I miss you and that I am doing my best to make things right for you. I can't say anything more.

MIA: Can you, please, just tell me if there is something special about me that interests "them," beyond the realm of the normal red blooded American citizen?

JORDAN: I think we both know the answer to that and it's, yes!

MIA: Are there more than just one intelligence agency interested in me?

JORDAN: You know that everything we're saying, our whole conversation, is being listened to, outside of us, and I don't want to say or give away my level of knowledge... not quite yet, so I can't answer all of your questions.

MIA: I just wish that I could believe that all of your feelings for me are true and that I'm not just some kind of subject or experiment.

JORDAN: You can rest assured nothing has changed and that I love you very much.

MIA: I sincerely hope so, since that is one of the few things that is keeping me going through this process. More than anything, I want to understand what this is about. Are you okay, where you are at?

JORDAN: They are treating me fine, I'm getting V.I.P. treatment. I was met at the plane by a guy in civilian clothes, driving a Ford Explorer. I was expecting a military person to take me here, but apparently they're not. I was brought to a reception room where a guy was watching the "X-Files." He handed me a cup of coffee and asked me what I think about the scene that was on the TV screen. It was where the FBI agent, Mulder, was near death after his encounter with the alien, and said that it was worth his dying, since he found out that all he believed was true, was true. The guy said to me, "that's a crock of shit, isn't it? What do you think? Would it be worth dying for?" When I heard him ask me this, I wondered if this was just a coincidence or whether this had been staged for my benefit or "theirs." I answered him that I didn't know whether it was worth it or not. Today they introduced me to a woman who took me around the facility. I have seen a lot of things here that don't look like they have anything to do with protecting us from the Russians. Tomorrow is supposed to be the big day for me here. My life will never be the same after seeing what I have seen. I don't like it.

MIA: Are they allowing you to carry your gun there? Are you being driven around in vehicles with blacked out windows? Are you being followed at all?

JORDAN: They have allowed me to carry my firearm. No, they are generally ordinary four-wheel vehicles that we travel in. I don't get the feeling that they are surveilling me, when I am on my own time. I can't say more about this.

MIA: I never heard from any of those people you thought were

113

going to contact me, or received anything. I am as much in the dark as I ever was. Is it okay for you to phone me?

JORDAN: Since the call got through, I guess it is okay. If they cut us off, I guess it is not.

MIA: Just more of the rat in the maze business. I want to understand my feelings about you. This feeling that you are bonded to me and are my alter ego or soul mate is so pervasive within me, but there is another part of me that is having a hard time believing that. I have always thought that I was the kind of person who is in control of her life and destiny. I'm beginning to feel that is a fallacy and I never really had that opportunity, ever! That whomever was determining my destiny, had it all planned out somehow and that is *not* very empowering. The only thing that *is* empowering to me is that I am able to endure this and still retain my sanity. I have found out how strong I am and that makes me feel good about myself. I don't think there are very many people who could have gone through this and stayed intact emotionally. How about you, how are you emotionally?

JORDAN: I have had a lot of ah.... I don't know how to put this ... expanding, outstanding experiences going on here... lots of learning experiences. I thought that you were going through the same process but I guess that you're not. I also find that interesting.

MIA: Nope, I am not going through anything except frustration. At least, I am not going through anything on a conscious level. Did you ever get the two letters that I wrote to you?

JORDAN: No, I haven't been to the office more than twice, since I last saw you.

MIA: Perhaps, they're holding them from you. The first one, you should have receive two days after you regression. You mentioned that a woman is showing you around Nellis, maybe she is the same woman who has struck again in my life. Do you think she is the same one who called the UFO hot line and left that strange message about me and also left that stupid address thing on my answering machine?

JORDAN: I don't think she is the person.

MIA: If that's the case, I guess we have a lot of people involved in this. Are you still going to Dulce and Aztec, New Mexico?

JORDAN: We're headed there in a day or two.

MIA: When you asked me about that date, do you know more about that date now?

JORDAN: Yes, that's what I need to talk to you about, but not on the phone.

MIA: Do you have any information for me:

JORDAN: I believe I do, I want to verify something with you. It happens that I am being permitted to do so.

MIA: Do you like being a puppet?

JORDAN: Maybe so, but at the same time, if whoever is holding the strings is taking me where I want to go, it's not so bad.

MIA: Were you really a skeptic, before you met me?

JORDAN: You mean about this stuff? Absolutely, that's why it is so hard.

MIA: Well, it's a happy little accident then.

JORDAN: I think a lot is going to happen shortly. I'm seeing it and I think we are going to play a part.

MIA: At this point in time, I wouldn't be surprised. I would hope that this is for some good reason and that those who serve the public, that are abusing our constitutional rights of privacy, are doing this for a real purpose!

JORDAN: I think that is what's going on.

MIA: I guess this is a major paradigm shift for you.

JORDAN: Yes.

MIA: I hope that I don't have to be afraid to see you to find out just what is going on here.

JORDAN: Why do you continue to say that?

MIA: Because I don't know where you're really at? You have

done a lot of lying to me but then again perhaps you weren't functioning at your optimum brain power. There is a chance, in my opinion, that you are being master minded, being manipulated. I say that because for a special agent, you made a lot of blatant mistakes.

(Jordan didn't seem to know what I was talking about. I went through my check list of his activities, the listening device in my smoke detector, the phone call to the busy signal, the supposed check out of my car for bugs and his regression by a therapist who says he doesn't do regressions.)

JORDAN: Wow, I am really impressed. There is nothing I can say ...these people are good. I don't know how they accomplished the busy signal thing. I was talking to the squad secretary and my supervisor. Inserting a busy signal, while someone is talking is impossible. As far as Dr. Weeks goes, I was referred by Yvonne Jones, a ufologist/hypnotist from the west coast who referred me to Barbara Martinez, a local ufologist, who works with Dr. Weeks.

MIA: I never heard of Barbara Martinez, she must not be very active in local UFO stuff. I did hear of Yvonne Jones but haven't met her. How would the powers that be know that I would check up on all of these things you did?

JORDAN: Exactly, that was their assumption that you *would* and that this would impugn my credibility. I honestly did, and found those things you indicated.

(I then discussed how I knew exactly what he was going to say about the abstract photograph in my apartment and how his saying exactly what I thought he would say added to the enigma about him and me.)

JORDAN: You can believe what you want.
MIA: Why would "they" want to discredit you?
JORDAN: It is a big chess game and they are playing four or five moves ahead of me. The options and planning is so far ahead.

MIA: Obviously, we're at a point where they are leading us in some direction.

JORDAN: They want to lead us away from each other.

MIA: But why then did they connect us at all? It doesn't make sense, unless there are two different groups or sponsors of the game. I want some definitive information from you, not just talk.

JORDAN: I understand.

MIA: Did you really find that code on your windshield?

JORDAN: I know what it says but I don't know if it is true or not.

MIA: Were you really that disturbed after your regression or did you not want to see me, because you knew I would hammer you with my disbelief and subject you to my inquisition?

JORDAN: I was upset.

MIA: But you blamed it on me!

JORDAN: No, I didn't.

MIA: You absolutely did! First you blamed "them" then you blamed me. That is what my first letter to you is about. If you disappear, I will never understand my feelings about you. That is why I have to have something factual, that I can relate to. Otherwise, your credibility is still on the negative side of the ledger.

(I proceeded to tell him about the phone call with the address in Miami, Dade County, Florida.)

JORDAN: That address is in the middle of nowhere. There is a jail on that street. I wouldn't go there, if I were you. It could be dangerous.

MIA: It could be or just another red herring, more crumbs thrown out for Hansel and Gretel on the way to the gingerbread house. You said that you were my protector.

JORDAN: A lot of good, I have done.

MIA: I pray that you won't let me down. Since you have come into my life, darling, you've added a whole new

117

dimension that has gone above and beyond. I hope that "they" will allow us to be privy to their machinations.

JORDAN: I am beginning to believe the wildest theories. What was reality to me is *not* reality any more. That I know for sure. I feel like we have done this before, that this is all rehearsed. I feel like other people are playing the roles but that I'm saying the lines! Like I have rehearsed with other people.

MIA: Remember months ago, when you had the deja vu about my African safari story. Was that a true feeling?

JORDAN: Absolutely. Mia, I am totally loyal to you. I felt that deja vu big time! The way I feel about you is so strong that I want only to be with you. I am willing to give up my family for us to be together.

MIA: I think that you should clarify the issues at home before you start talking that way. What about the question of your wife's being with the A.F.O.S.I. officer, was that true?

JORDAN: Yes, that is where this all started and that is why I hate her so much and could never recoup our relationship.

This conversation with Jordan, putatively while he was at the Nellis facility, continued for an hour and a half. His voice again seemed depressed and agitated. Not much else was said that was substantive, rather it seemed that he just wanted to talk to me in general.)

Four days separated the Nellis phone call from the next one. I was entertaining a childhood friend I hadn't seen for years when Jordan phoned at 4:15 P.M.)

MIA: Where are you?

JORDAN: I would rather not say. How are you doing?

MIA: Okay. Is there any chance that you could call me later, I have a visitor here from out of town and I can't speak for any length of time. Is there something you can tell me now?

118

JORDAN: It is something that will take time. It requires a lot of time. I'll try to call back.

MIA: No matter what this looks like to me, I want you to try to explain what has happened. I don't want to close myself off to you.

JORDAN: Will you tell me one thing?

MIA: Sure.

JORDAN: Could you tell me the nature of your feelings about me?

MIA: Do you mean, do I love you? Of course, I do. But you know the other things that are happening

JORDAN: Do you love me romantically, sexually or platonically?

MIA: Well, it certainly isn't platonic, of course not...But with all of these other things going on, I am afraid of making a specific commitment. I don't want to be hurt over or by this.

JORDAN: That tells me a lot, as far as what I need to do.

MIA: If I felt anything other than that, after all I told you the other night, I wouldn't even talk to you.

JORDAN: I am at least grateful that you are able to tell me that, because now I know what I have to do.

MIA: Whatever it is, it has to be your own decision. There are so many unanswered questions here.

JORDAN: I think I am going to be permitted to answer most of your questions.

MIA: I don't want to be manipulated into something that is going to be injurious to me.

JORDAN: You won't be. I think you are going to be given an opportunity to at least hear what I understand to be the truth. You might not like it but I can tell you the truth, at least, as much of the truth as I understand it.

MIA: I wish someone would! I am tired of people telling me lies or playing around with my head. No matter what is said, I think I can accept it or reject it, if I choose. In this case, something is better than nothing!

JORDAN: The problem is I don't know how much of what I think is the truth, may turn out to be a lie.

MIA: I can understand the nature of duplicity here.

JORDAN: My time here is over today and I'm going up north. I
         have a lot to tell you. Hopefully, the next stage of this
         trip is not an extension of this phase. One thing you
         ought to do is just keep your eyes open.

MIA: I keep hearing you saying that. Am I supposed to be
         frightened of that?

JORDAN: No, it is not to scare you. It is just so that you not miss
         anything.

Once again, Jordan cryptically made mysterious but unsubstantial statements. At this point, like Shakespeare said, "It is much ado about nothing." When, if ever, would he appear with something that was meaningful to me. My patience was ebbing and wearing thin at best.

MORE STUPID GAMES?

Jordan never phoned back in the evening as he said he would. I detested the waiting, always fearing that if I took a shower or went to the bank, I would miss his call. That, in itself, was the cruelest of manipulations. I continuously left messages on his voice mail and beeper but to no avail. I decided to mount a more aggressive attack. After a week, I phoned the FBI office and asked the squad secretary when he would be back in the office. She took my name and put me on hold. Within a few seconds, she returned with the news that he "would probably be back tomorrow." At least, that was something! In less then five minutes, Jordan phoned. Apparently, he was back. He sounded calm and emotionally intact. Without any preliminaries or defensive statements he said, "I heard you called the office." "Yes, I replied, "I was concerned about you, your voice on your outgoing voice mail message sounded terribly depressed.; "I don't know why you say that, I am okay and I know you are. Very soon, you will know everything. Read page 324 for something important that you need to know."

"This is crap, more ridiculous games," I muttered to myself. "I don't know what you are talking about!"

Jordan said assertively, "I know you do, I have every confidence you do, just think about it." I wanted to scream! I couldn't believe that he was doing this again. "No, I don't."

Speaking slowly and precisely, Jordan continued. "Remember the book you loaned me. It has something very important in it on page 324. I can't say anything more definitive because 'they' will come and try to get it. Just think about it, I know you will understand. I know you can do it. I have to go now."

"But will I see you soon?" I demanded.

121

"Yes, very soon."

I was totally enraged! I had absolutely no idea what page 324 might mean. Three months before, I had loaned him some books. I could remember only the names of four. I kind of remembered that there was one more, but I couldn't think of it! The first book I loaned him was Tim Good's *Above Top Secret*. I purposefully gave him that because it had a lot of good data on government and intelligence agency coverups. I rushed to page 324 and found a chapter on Peru and an attempt to shoot down a spacecraft. I had been to Peru but that was years after the incident in the book. I checked the other three books but they had less than 300 pages. Now, I was really going berserk. I was afraid that I was going to miss something important and that Jordan was overestimating my abilities, whatever they were, and then I was absolutely furious that he was still playing these games on me.

My fury was just for a few instants, when all of a sudden, like the proverbial light bulb going on over my head, I walked to my over loaded and jam packed bookshelf and picked up a book that I had never really read but just browsed through. I held the book and knew that was the other book I had forgotten about and whose name and author, I could not remember. If someone would have given me a million dollars, I couldn't have remembered! I bought *Phenomenon* by Spencer and Evans when I visited Glastonbury, England four years before. I had loaned it to Jordan when he asked me if I knew anything about the Bentwaters incident in England. It had over 300 pages in it. I opened it up to page 324 and the chapter was about light phenomenon, especially in ancient sites. In the middle of the page, in between paragraphs, was something hand printed in pencil. The printing looked like Jordan's. It said *"meet me at Sage at 12 noon tomorrow!"* My first thought was that Jordan had written this as a memo to himself, about a meal we planned to have together in the past. The first restaurant we went to as friends, after our initial meeting at Denny's about crazy Mark, was the Sage restaurant. We had subsequently gone to many other places to eat.

That didn't make sense. Jordan is not the type to write in other people's books. Then the light bulb went on over my head a second time, only this time it blinded me! "Oh, my God," I thought, "this is what he told me to look for! This is a message for me to meet him tomorrow at the Sage restaurant! But how did it ever get in the book and, more importantly, *when* did he put it in the book?" I was in total shock. He returned the book to me over two months before and as far as I knew never again had it in his hot little hands! That meant that this whole business of superspy gaming was pre- planned and he was anticipating this tete-a-tete well in advance, or even more sinister, he might have come into my apartment and put it in recently! All of a sudden, my mental and emotional computer went on overload! A torrent of emotions was raining on and through me. I was in an emotional tunnel of darkness.

First of all, how did I know to go directly to that book? I had put it completely out of my mind. It scared me because Jordan was so certain I would know what book he meant and even after I threw my hands up and cried to the "All Seeing Eye" that I didn't have the foggiest notion what book he was talking about, I seconds later walked to the exact book! I actually screamed out loud, asking how I did that? What forces at work turned me in the proper direction? Even more baffling was the secret notation within. The only solace was that in less than 24 hours, I would know everything. Yeah, right, if Jordan was true to form, that was iffy, at best, but so far, it was my best shot!

I felt that after all that had ensued there might be a chance that this was a set-up, a way to get me off kilter and unprepared for possible danger. I didn't think that this was paranoia, I thought it was good, rational thinking, considering episodes of the past. I discussed this with my friends and confidantes, Susan and Cath, who have been with me every step of this crazy and momentous journey. We decided that they would also lunch at Sage, and they would monitor the progress of our meeting. We had a signal pre-arranged, in case I was in trouble or anticipating trouble. Susan and Cath had been ordained my cosmic bodyguards. They both just happened to have expertise in karate, especially Cath, and we knew that we had a whole bevy of help available on the unseen levels, so we were all set for action and raring to go!

As it happened, my cynicism about Jordan's following through with our luncheon meeting was correct. Amazingly, I slept well the night before the meeting. I fell asleep quickly and deeply. I had feared that the anxiety produced by the anticipation would cast me back into the throes of my insomnia of the past. It did not. I couldn't wait for the moment we would hug and I would find out just what the hell was going on! Jordan had been promising me results for over three months and I was now close to the truth!

An hour before we were to meet, the phone rang. It was Jordan speaking softly and tremulously. He said that he was sorry but he couldn't come to meet me! I was very upset and said so. I was frustrated again! How could he do this to me! He sounded so self assured and cool the day before. Now he had that pathetic tone in his voice again. I don't mean to sound uncharitable, but he was pissing me off! Jordan said that I shouldn't be upset because I would have something in hand today and he would make sure of it. Since I had no choice, I decided to make the best of it and stayed home and caught up on my paperwork. I wanted to be home, just in case he decided to make good on his promise. I was not very optimistic about that. I phoned my friends Susan and Cath and told them what had transpired and that the cosmic bodyguards did not have to show up for work today. They were as disappointed as I was!

At 4:20 P.M., my doorman phoned to tell me that I had a Federal Express package. I knew in my gut that was what I had been waiting for. I went to retrieve it. Sure enough it was a Federal Express package but it didn't have the normal name and sender documentation on it. It was wrapped tightly in a plastic sleeve with my name typewritten 20 times on blue paper. The doorman said that a guy left it who was driving a regular car, not a Federal Express

124

truck. I knew that was Jordan, the little chicken shit! I was still mad, so I thought this better be good stuff! Good wasn't the word for it.

Following is the actual report and follow up phone call. It was much more than I had bargained for. As an added surprise, Jordan had included a monogrammed patch from Nellis AFB that said "Dreamland, Area 51 Police, Groom Dry Lake Test Facility." Some souvenir that is!

"As they say, the truth is out there, but can we recognize it when we see it?"

–Jordan Perez
FBI Special Agent

THE REPORT

This is the report Jordan delivered to me the last day we spoke early in 1995. Upon scanning the report, what became most obvious to me were the continuity of misspelled words and strange transition of the lower case letter "n", as seen in the word "OnLY." The misspellings are very odd considering Jordan's high level of literacy and his position that requires written reports. Thus, the misspellings of a simple word like "INFORMATION" seemed strange, indeed, since that word would be in every report written for the FBI. Jordan had been on full academic scholarship for four years at Duke University, therefore he should have been able to spell these simple words. Since he continually misspelled the words as INFORMETION, INVOLVD, BELEEF, INVOLVS, INVOLVD, INVOLV, WOULDE and DISINFORMETION, this may be more than just normal typing mistakes. The continual appearance of the lower case "n" in the word OnLY is unexplainable in normal technical terms. Other "O-N" combinations do not appear similarly. My conclusion, and that of others who have also sensed the strangeness of this, is that there might be some subliminal or coded message within the text of the report. James Bondish as it sounds, I believe it is a real possibility, although I have no idea what it might be.

The information given by Jordan is not sanctioned by me as being accurate. Just as Jordan says, even he was not sure that he wasn't being used to give this out as disinformation. It is interesting, though, that the information is quite similar to that given to me by Panda in 1989. Judge for yourself and pray that it is a lie!

THE ACCOUNT I AM ABOUT TO GIVE IS BASED ON MY EXPERIENCES AND THE PERSONAL INVESTIGATION I HAVE CONDUCTED OVER THE PAST SEVERAL MONTHS. I CANNOT, AND IN A FEW INSTANCES I AM SPECIFICALLY PREVENTED FROM, REVEALING WHERE

AND WHEN I ACQUIRED ALL OF THE INFORMETION I WILL PRESENT. I WILL STATE THAT ON TWO OCCASIONS I DELIBERATELY VIOLATED SECURE AREAS AND ON ONE OCCASION PARTICIPATED IN A RATHER UNORTHODOX FIELD INTERROGATION IN ORDER TO OBTAIN DATA. I ALSO DELIBERATELY INVOLVD SEVERAL INNOCENT PARTIES WHO, UNFORTUNATELY, MAY BECOME SUBJECT TO PENALTIES WHICH SHOULD RIGHTLY OnLY BE MINE. I HAVE LIED AND I HAVE DONE SO KNOWINGLY AND REPEATEDLY IN ORDER TO ELICIT INFORMETION AND IN ORDER TO CAUSE GOVERNMENTAL ENTITIES TO REACT TO MOVES THEY OnLY THOUGHT I WAS MAKING. I FREELY ADMIT THAT I AM GUILTY OF ABUSING THE POWER OF MY OFFICE AND MY CREDENTIALS. I DO NOT APOLOGIZE FOR THIS, BUT I MAY SOON BECOME SUBJECT TO THE ADMINISTRATIVE OR LEGAL CONSEQUENCES OF THESE ACTIONS. SO BE IT. AT THE RISK OF APPEARING MELODRAMATIC, MORE DIRE CONSEQUENCES MAY ALSO FOLLOW. I CAN OnLY STATE THAT, IF RESISTANCE IS WITHIN MY POWER, I WILL NOT GO QUIETLY.

I AM TAKING THE PRECAUTION OF DISSEMINATING THIS ACCOUNT (BOTH ON PAPER AND ON DISKS) TO SEVERAL INDIVIDUALS SOLELY AND EXCLUSIVELY AS A MEANS OF FORESTALLING THE DESIRABILITY OF ATTEMPTING TO SILENCE ME ENTIRELY. IT IS MY BELEEF THAT REVENGE WOULDE NOT BE THE GOAL OF THOSE OPPOSED TO MY EFFORTS, BUT, RATHER, THE PREVENTION OF THE LEAKAGE OF INFORMETION. IF THIS IS A **FAIT ACCOMPLI,** THEN PERHAPS I WILL BE LEFT RELATIVELY UNDISTURBED AS I HAVE NO INTENTION OR DESIRE OF PURSUING THESE MATTERS FURTHER. I HOPE THAT I WILL BE LEFT ALONE FROM NOW ON.

AS A COPY OF THIS REPORT IS NOW IN YOUR HANDS, I URGE YOU TO MAKE NUMEROUS ADDITIONAL COPIES AND TO PLACE THESE COPIES IN SEVERAL LOCATIONS

TO AVERT EFFORTS AT RECOVERING ALL OF THEM. LET ME BE VERY CLEAR: IF MY OPPONENTS ARE ABLE TO RECOVER ALL THE COPIES OF MY REPORT, THEN THEY WILL INEVITABLY TURN TO THE MATTER OF MUZZLING ME AS THE OnLY REMAINING THREAT TO SECURITY THEY FACE. HOWEVER, IF YOU ARE CONCERNED WITH THE CLASSIFIED NATURE OF THIS DATA AND DECIDE TO DESTROY THIS INFORMETION INSTEAD, I WOULDE ASK THAT YOU DO SO THOROUGHLY SO THAT IT DOES NOT FALL INTO THE HANDS OF A THIRD PARTY. I ALSO STRONGLY PLEAD THAT MY NAME, IF IT IS KNOWN TO YOU, NOT BE DISSEMINATED AS, IF AN ATTEMPT IS MADE TO VERIFY MY AUTHORSHIP, I MUST MAKE VEHEMENT DENIALS. HOWEVER, I HOPE THAT THE INFORMETION CONTAINED IN THIS DOCUMENT WILL BE OF SOME POSITIVE USE TO YOU AS I, AND OTHERS SYMPATHETIC TO MY EFFORTS, HAVE TAKEN TREMENDOUS RISKS IN OBTAINING IT.

I AM FIRMLY CONVINCED THAT I WAS ABDUCTED BY NON-HUMAN SENTIENT BEINGS AT THE AGE OF SIX (████) AND THAT MY NEWLY SURFACED (THANKS TO DR.████ ████ ██ ██ ██████ █████ ██ ███████ RECOLLECTIONS REGARDING THIS INCIDENT ARE RELIABLE AND REAL. THESE BEINGS STRONGLY RESEMBLED THE GREY "BREEDERS" I WILL DESCRIBE BELOW, BUT, AS I REMEMBER NO "WORKERS," I CANNOT BE ENTIRELY SURE OF WHAT THESE BEINGS REPRESENTED. HOWEVER, I HAVE INCONTROVERTIBLE EVIDENCE THAT THE ALIENS WHO ABDUCTED ME AS A CHILD HAVE MAINTAINED AN INTEREST, IF NOT AN INFLUENCE IN MY LIFE. THIS EVIDENCE HAS ALREADY BEEN PROVIDED TO THOSE WHO COULD MAKE BEST USE OF IT.

I HAVE BEEN ADVISED THAT MY OWN BIRTH WAS AS A RESULT OF GENETIC INTERVENTION BY THESE BEINGS. I RECALL THAT MY MOTHER AND FATHER ALWAYS HAD STATED THAT MY MOTHER WAS DIAGNOSED AS HAVING

BEEN INCAPABLE OF BECOMING PREGNANT AND THAT I WAS, THEREFORE A "SURPRISE." FURTHER.MORE, AFTER MY BIRTH, MEDICAL EXAMINATION REVEALED THAT MY MOTHER SHOULD NEVER HAVE BEEN ABLE TO CONCEIVE ME. I DO KNOW THAT MY PARENTS WERE NEVER ABLE TO CONCEIVE AGAIN, DESPITE EFFORTS TO DO SO. MY SOURCE ADVISED ME THAT A DONOR EGG WAS OBTAINED FROM ANOTHER WOMAN AND HER GENETIC MATERIAL ERASED FROM THE EGG. MY MOTHER'S GENETIC MATERIAL WAS THEN SUBSTITUTED AND THE EGG FERTILIZED BY MY FATHER'S SPERM. I MUST THEREFORE CONCLUDE THAT AT LEAST MY MOTHER WAS ALSO AN ABDUCTEE.

WHY OR EVEN IF MY BIRTH WAS OF PARTICULAR INTEREST TO THE ALIENS IS UNKNOWN TO ME. HOWEVER, I HAVE REASONS TO BELIEVE THAT I WAS CONDITIONED TO SEEK EMPLOYMENT WITH THE UNITED STATES GOVERNMENT.

VERY RECENTLY I WAS DELIBERATELY AND RATHER ARTFULLY PLACED IN CONTACT WITH THE WOMAN WHOSE EGG WAS USED TO CONCEIVE ME. I UNDERSTAND THAT THIS WAS AN ALIEN EFFORT TO DETERMINE THE INFLUENCE OF NON-TANGIBLE (ENERGY FIELDS?) FACTORS IN HUMAN EMOTIONAL INTERACTION. WHEN THE RESULT WAS, IF THIS IS THE RIGHT TERM, INCESTUOUS, CERTAIN GOVERNMENT AGENTS ATTEMPTED TO CONTRADICT THE EBE EFFORTS TO FURTHER THE RELATIONSHIP. I AM DEEPLY GRATEFUL FOR THEIR EFFORTS AS I AM CONVINCED THAT THE OnLY THING WE MAY HAVE LEFT TO SAVE AS A RACE IS OUR HUMAN DIGNITY. THESE AGENTS BELONG TO AN AGENCY KNOWN AS THE DEPARTMENT OF THE NAVY.

THE DEPARTMENT OF THE NAVY (DON) IS A RELATIVELY SMALL CLANDESTINE AGENCY WITHIN THE UNITED STATES GOVERNMENT WHICH IS STAFFED PRIMARILY

BY INDIVIDUALS OSTENSIBLY RECRUITED AND TRAINED BY OTHER ELEMENTS OF THE FEDERAL GOVERNMENT. IT IS FUNDED AND SUPPORTED BY THE NATIONAL SECURITY AGENCY AND IS ALSO AUGMENTED BY VERY TIGHTLY COMPARTMENTALIZED ELEMENTS OF THE ARMED FORCES (I BELIEVE THAT BLUE LIGHT IS THE ORIGINAL CODE NAME FOR THE U.S. ARMY'S EBE-RELATED UNIT, WHICH WAS ALLEGEDLY CONVERTED INTO THE NOW FAMOUS COUNTER-TERRORIST DELTA FORCE) AND OTHER GOVERNMENTAL ORGANIZATIONS. (APPROXIMATELY ONE THIRD OF HEADQUARTERS DIVISION FIVE OF THE FBI IS ACTUALLY A FRONT FOR THE DON, WITH AT LEAST SEVERAL DOZEN HQ SPECIAL AGENTS DEVOTED TO THESE MATTERS.)

DESPITE ITS NAME, THE DEPARTMENT OF THE NAVY HAS VERY LITTLE TO DO WITH THE ACTUAL UNITED STATES NAVY. ITS EXISTENCE IS KNOWN OnLY BY AN EXTREMELY LIMITED NUMBER OF INDIVIDUALS AND ITS ACTUAL MISSION IS KNOWN BY AN EVEN SMALLER NUMBER. THE DEPARTMENT OF THE NAVY'S PURPOSE IS TO DIRECT EXTRAORDINARILY UNIQUE ACTIVITIES IN RELATION TO EXTRATERRESTRIAL BIOLOGICAL ENTITIES (EBEs) AND ALIEN TECHNOLOGIES WITH THE PRIMARY MISSIONS OF COMBATTING ALIEN OPERATIONS WHICH ARE HOSTILE TO THE NATIONAL SECURITY OF THE UNITED STATES, GATHERING AND EXPLOITING ALIEN TECHNOLOGIES, AND THE CONTAINMENT OF PREMATURE DISCLOSURE OF INFORMETION REGARDING THE PRESENCE AND INTENTIONS OF EBEs.

IRONICALLY, WHILE NASA SUPPOSEDLY SUSPECTS OR KNOWS THAT THERE ARE ALIEN ARTIFACTS ON THE MOON AND IS SUPPRESSING DATA ON THIS SUBJECT (THIS WAS OnLY MENTIONED TO ME AS AN ASIDE), NASA HAS NO OFFICIAL ROLE IN DEALING WITH EBEs. THIS IS NOT TO SAY, HOWEVER, THAT NASA, LIKE THE

REST OF THE OVERT GOVERNMENT) IS NOT PENETRATED BY SOME DON PERSONNEL. FURTHERMORE, THE DON HAS HAD SOME INTERACTION WITH NASA PERSONNEL WHO ACCIDENTALLY BECAME AWARE OF TOO MUCH. I WAS ALSO TOLD THAT AN OVERT NASA EFFORT TO DETECT NON-HUMAN CIVILIZATIONS WAS DOOMED TO FAILURE, BUT I WAS NOT TOLD WHY.

THE MOST CLASSIFIED ASPECT OF THE DON's ACTIVITIES IS THAT IT OPERATES IN CONCERT WITH EBEs OF A PARTICULAR SPECIES COMMOnLY IDENTIFIED AS" GREYS." GREYS ARE SURPRISINGLY ANATOMICALLY AND GENETICALLY SIMILAR TO HUMANS. HOWEVER, THE GREYS ARE DIVIDED INTO QUASI-SENTIENT ASEXUAL "WORKERS" WHO ARE, ON AVERAGE, FOUR FEET TALL, AND DOMINANT — BREEDERS- WHO HAVE LARGER EYES AND ARE, ON AVERAGE, ALMOST SIX FEET TALL. BOTH TYPES OF GREYS HAVE FOUR FINGERS ON EACH HAND. THIS ALLIANCE EMERGED FROM THE SUCCESSFUL RECOVERY OF SEVERAL APPARENTLY CRASHED GREY CRAFT AND THE SUBSEQUENT ESTABLISHMENT OF RELATIONS WITH THIS SPECIES. THIS RELATIONSHIP IS, FOR THE TIME BEING, CLANDESTINE BY THE CHOICE OF BOTH PARTIES AND INVOLVS A VERY LIMITED AND GRADUAL SHARING OF INFORMETION AND TECHNOLOGY BY THE GREYS. SEVERAL UNDERGROUND AND UNDERWATER (AT LEAST ONE UNDERWATER FACILITY IS OFF THE FLORIDA COAST AND HAS A TERRESTRIAL ACCESS POINT IN SOUTH MIAMI I AND MY UNSEEN COHORTS HAVE ALREADY PROVIDED THE ADDRESS TO SEVERAL INDIVIDUALS) FACILITIES HAVE BEEN ESTABLISHED TO SUPPORT THESE ACTIVITIES AND TO PROVIDE A SUITABLE HABITAT FOR THE EBEs INVOLVD. PARENTHETICALLY, THE MULTIPLE PROTECTIVE SHELTER (MPS) ICBM BASING SCHEME PROPOSED IN THE EARLY 1980s WAS ACTUALLY A COVER FOR THE CONSTRUCTION OF 4,600

EBE RELATED BUNKERS; THIS SCHEME WAS POLITICALLY UNTENABLE AND HAD TO BE ABANDONED, BUT WOULDE HAVE ALLOWED FOR A TREMENDOUS EXPANSION OF ALIEN ACTIVITY IN THE UNITED STATES.

THERE IS PRESENTLY AN ON-GOING COMPETITION FOR INFLUENCE, IF NOT OUTRIGHT DOMINION, OVER THE EARTH AND ITS LIFE FORMS BETWEEN THE GREYS AND ANOTHER SPECIES COMMONLY IDENTIFIED AS "REPTILIANS" OR "LIZARDS." IT IS SUPPOSEDLY KNOWN THAT THIS SPECIES OMINOUSLY REFERS TO ITSELF AS "EARTHLINGS." I WAS SHOWN EXTREMELY CLEAR PHOTOGRAPHS OF EXAMPLES OF THIS SPECIES. REPTILIANS APPEARED HUMAN SIZED, HAD THREE TOES ON EACH FOOT, AND HAD NOTABLY LARGE MOUTHS WITH PROMINENT TEETH. THIS COMPETITION HAS RESULTED IN THE DESTRUCTION OF SEVERAL CRAFT BOTH WITHIN AND WITHOUT THE EARTH'S ATMOSPHERE AND HAS PROMPTED THE DEVELOPMENT OF CERTAIN ASPECTS OF THE UNITED STATES' STRATEGIC NUCLEAR FORCES AS WELL AS THE STRATEGIC DEFENSE INITIATIVE (SDI).

THE DON CONDUCTS EXTENSIVE SURVEILLANCE AND MONITORING OF INDIVIDUALS WHO ARE UNKNOWINGLY PART OF AN ON-GOING PROGRAM OF MANIPULATION OF THE HUMAN GENE POOL BY COMPETING NON-HUMAN ALLIANCES. THIS MANIPULATION WAS INITIALLY UNDERTAKEN BY THE REPTILIAN SPECIES, APPARENTLY TO DIRECT CERTAIN HUMAN TRAITS IN WAYS FAVORABLE TO THIS SPECIES. IN ADDITION REPTILIAN EFFORTS HAVE APPARENTLY FOCUSED ON THE UNITED STATES AND, TO A LESSER EXTENT, ITS ALLIES DUE TO THE UNITED STATES, LEADING ECONOMIC AN MILITARY ROLE WORLDWIDE. STUDY HAS DETERMINED THAT THE REPTILIA EFFORTS IN THIS AREA AMOUNT TO LONG-TERM GENETIC SABOTAGE. THE REPTILIANS HAVE

ALSO BEEN RESPONSIBLE FOR LIVESTOCK MUTILATIONS AND SOME HUMAN ABDUCTIONS FOR OTHER NEGATIVE PURPOSES. STRANGELY, THERE IS SOME HISTORICAL CONNECTION BETWEEN THE REPTILIANS AND THE DEVELOPMENT OF NUCLEAR ENERGY, BUT THIS WAS OnLY HINTED AT BY MY SOURCES.

HOWEVER, THE GREY ABDUCTIONS OF HUMAN SUBJECTS HAVE BEEN AIMED PRIMARILY AT THE CREATION OF A PARTICULARLY HARDY ELITE HYBRID SPECIES WHICH WOULDE BE INITIALLY CAPABLE OF FORMING THE LEADING ELEMENTS OF GREY SOCIETY AS WELL AS SERVING A SIMILAR ROLE IN HUMAN SOCIETY. THE HYBRIDS WOULDE ALSO BE ABLE TO SURVIVE DESTRUCTIVE ECOLOGICAL CHANGES THAT THE GREYS BELIEVE WILL TAKE PLACE ON EARTH IN THE NEAR FUTURE. THE HYBRIDS I SAW HAD THE APPEARANCE OF HUMAN CHILDREN WITH UNUSUALLY LARGE HEADS AND EYES AS WELL AS FINE, SPARSE HAIR. EARLIER, AND UNSATISFACTORY, EFFORTS SUPPOSEDLY RESULTED IN A MORE FETUS-LIKE APPEARANCE ON THE PART OF THESE HYBRIDS.ALL HYBRIDS THUS FAR PRODUCED ARE STERILE AND UNABLE TO REPRODUCE. HOWEVER, ALL HYBRIDS HAVE WELL-DEVELOPED DIGESTIVE SYSTEMS.

THE IMPLANTATION OF BIO-ELECTRONIC (?) DEVICES IN PROXIMITY THE CENTRAL NERVOUS SYSTEMS OF ABDUCTEES HAS BEEN UTILIZED BY THE GREYS FOR SEVERAL COMPLEMENTARY PURPOSES, BUT THE MOST IMPORTANT REASON IS IN ORDER TO HAVE A MORE RELIABLE MEANS OF MIND CONTROL AT A DISTANCE. SUCH CONTROL IS NECESSARY IN ORDER TO FACILITATE FOLLOW-UP ABDUCTIONS AND IN ORDER TO PREVENT MALE SUBJECTS FROM HAVING VOLUNTARY STERILIZATIONS.

THE DON HAS TAKEN A VERY ACTIVE, AND LARGELY SUCCESSFUL, ROLE IN COUNTERING REPTILIAN GENETIC SABOTAGE. HOWEVER, IT IS TASKED WITH TAKING NO ROLE OTHER THAN OBSERVATION OF GREY ABDUCTIONS, TO THE EXTENT PERMITTED BY THIS GROUP. THIS FACT RESULTS IN MORALE AND CONTROL PROBLEMS WITHIN THIS ORGANIZATION. (THESE MORALE PROBLEMS HAVE BENEFITTED ME ON MORE THAN ONE OCCASION AND I STRONGLY BELIEVE THAT GOVERNMENTAL COLLUSION IN THESE ACTIVITIES IS UNCONSTITUTIONAL.) THE GREYS, LIKE THE REPTILIANS, ROUTINELY ATTEMPT TO ERASE THE ABDUCTION RELATED MEMORIES OF THOSE THEY ABDUCT.

ONE OF THE MOST EXOTIC ELEMENTS OF GREY TECHNOLOGY EXPLOITED BY THE DON INVOLV THE DELIBERATE REGULATION OF HUMAN BRAIN ACTIVITY INCLUDING TELEPATHIC COMMUNICATION, AND PROPULSIVE TECHNOLOGIES WHICH INCLUDE TEMPORAL AND, FOR LACK OF A BETTER TERM, "TRANS- DIMENSIONAL" FACETS. GREY BIOTECHNOLOGIES ARE, PARADOXICALLY, FUNDAMENTALLY PRIMITIVE, BUT HAVE SOME EXTREMELY ADVANCED FACETS. PARTICULARLY SIGNIFICANT IS AN ABILITY TO MANUFACTURE SYNTHETIC BIOLOGICAL-MECHANICAL (?) ENTITIES. I WONDER IF THE SO-CALLED "MEN IN BLACK," IF THEY EXIST, ARE EXAMPLES OF THESE SORTS OF BEINGS.

THE CONDITIONING OF HUMAN BRAIN (TEMPORAL LOBE) ACTIVITY THROUGH ELECTROMAGNETIC AND ADVANCED MEDICAL TECHNIQUES IS USED BY THE DON IN TWO PRIMARY WAYS. THE MOST OBVIOUS IS IN THE ELIMINATION OF UNDESIRABLE MEMORIES FROM THE MINDS OF ACCIDENTAL WITNESSES TO GREY ABDUCTIONS; IN THIS CASE SOMEWHAT LESS RELIABLE ELECTROMAGNETIC MEANS MUST

NORMALLY UTILIZED BE UTILIZED. THE SECOND, BUT ACTUALLY MORE IMPORTANT USE FOR THIS FORM OF ADVANCED MIND CONTROL IS IN THE ELIMINATION OF UNDESIRABLE MEMORIES IN THE MINDS OF DON PERSONNEL AND OTHER KNOWING PARTICIPANTS IN GOVERNMENTAL ACTIVITIES RELATING TO EBEs. THIS LATTER FORM OF HYPNOSIS IS NORMALLY DONE THROUGH MORE RELIABLE MEDICAL OR CHEMICAL MEANS AND IS UTILIZED FOR TWO REASONS. THE FIRST REASON IS ACTUALLY IN THE INTERESTS OF THE MENTAL HEALTH OF CERTAIN INDIVIDUALS UNABLE TO SUCCESSFULLY COPE WITH THE UNIQUE PSYCHIC ENVIRONMENT SURROUNDING EBE ACTIVITIES. THE SECOND REASON TO UTILIZE VOLUNTARY MIND CONTROL ON GOVERNMENT PERSONNEL IS TO DELETE INFORMETION FROM THE MINDS OF THESE INDIVIDUALS WHICH COULD BE EXPLOITED BY EITHER TERRESTRIAL OR EXTRATERRESTRIAL FORCES.

DON FIELD AGENTS ARE THEREFORE OFTEN REFERRED TO AS "ZOMBIES" AND THEIR UNITS KNOWN AS "ZOMBIE SQUADS." VOLUNTARY ERASURES ARE FAR MORE RELIABLE THAN THOSE IMPOSED UPON THE SUBJECT, ESPECIALLY IF THEY HAVE A SURGICAL COMPONENT. HOWEVER, NEITHER TECHNIQUE IS ABSOLUTELY RELIABLE AND, THEREFORE, ALL "ERASURES" ALSO INCLUDE ELEMENTS OF DISINFORMETION IMPLANTED IN THE MIND OF THE TARGET WHICH WOULDE TEND TO RENDER ANY RECALLED ACCOUNTS CONTRADICTORY OR OTHERWISE UNRELIABLE. ZOMBIES ARE ALSO PROGRAMMED AGAINST ALLOWING THEMSELVES TO BE HYPNOTICALLY REGRESSED OR OTHERWISE MEDICALLY EXAMINED BY UNAUTHORIZED PERSONNEL. (THIS IS PARTICULARLY IMPORTANT AS MOST OF THESE AGENTS HAVE IMPLANTS PLACED IN THEIR BODIES BY THE GREYS TO SUPPOSEDLY MONITOR DON COMPLIANCE WITH MUTUAL

AGREEMENTS.) MORALE AMONG DON AGENTS IS
PARTICULARLY IMPAIRED BY THE OBVIOUS
DETERIORATION OF MENTAL AND EMOTIONAL
FUNCTION, AMONG THESE INDIVIDUALS AS A RESULT
OF THESE PSYCHIC INTERVENTIONS.

THE DEPARTMENT OF DEFENSE HAS BEEN A PRIMARY
BENEFICIARY OF ALIEN DERIVED TECHNOLOGIES
WHICH HAVE BEEN DISSEMINATED IN A LARGELY
COVERT MANNER. IN OTHER WORDS, EVEN THE
MILITARY ENGINEERING UNITS AND DEFENSE
CONTRACTORS INVOLVD HAVE BEEN UNAWARE OF
THE ACTUAL SOURCE OF KEY BREAKTHROUGH
TECHNOLOGIES SUPPLIED TO THEM. ALTHOUGH SOME
ALIEN TECHNOLOGIES HAVE BEEN ADOPTED
(SPECIFICALLY IN THE HYPERSONIC "███████" STEALTH
AND "████████ ███████" RECONNAISSANCE PROGRAMS)
AND SOME ANTI-GRAVITY DRIVE EXTRATERRESTRIAL
CRAFT ARE NOW BEING ACTUALLY DUPLICATED (I
WAS SHOWN ONE HOVERING ALMOST SILENTLY IN A
HANGAR AT A SUB-SITE OF THE NELLIS AIR FORCE
BASE AND CAN STATE THAT RUMORS REGARDING
SUCH CRAFT WERE ACCURATE), THE MOST
SIGNIFICANT ALIEN TECHNOLOGIES BEING ACQUIRED
ARE SUPPOSEDLY IN THE FIELD OF COMPUTING.
THESE TECHNOLOGIES HAVE PROVIDED THE UNITED
STATES WITH A SECRET, AND THEREFORE LARGELY
THEORETICAL, TECHNOLOGICAL SUPERIORITY OVER
ALL OTHER NATIONS. I WILL ALSO MENTION THAT I
WAS FLOWN ABOARD A HELICOPTER OF A DESIGN I
HAD NEVER BEFORE SEEN WHICH WAS NOT OnLY
UNBELIEVABLY QUIET, BUT COULD ALSO "CLOAK"
ITSELF FROM BEING SEEN BY THE NAKED EYE DURING
THE HOURS OF DARKNESS USING A SYSTEM WHICH
WAS REFERRED TO AS "███"

THERE IS ALSO AN ONGOING PROGRAM OF GRADUAL
SOCIETAL CONDITIONING TO ACCEPT THE EXISTENCE
AND INFLUENCE OF EBEs IN EARTH'S HISTORY. THIS IS

BEING UNDERTAKEN BY THE DON ON A WORLDWIDE BASIS IN ORDER TO PREVENT CULTURAL DISINTEGRATION IN THE EVENT OF AN UNCONTROLLABLE CATASTROPHIC LEAK OF INTELLIGENCE REGARDING THE ALIENS AND ALSO IN ORDER TO ALLOW FOR A MORE ORDERLY EVENTUAL UNVEILING OF THE TRUTH TO THE GENERAL PUBLIC. THE GREYS HAVE THEMSELVES PROMULGATED A DESIRE TO OVERTLY REVEAL THEMSELVES TO HUMAN SOCIETY BY THE TURN OF THE CENTURY.

ACCORDINGLY, THE DON CURRENTLY HAS PLANS TO EVENTUALLY STAGE A FIRST CONTACT EVENT SIMILAR TO THAT DEPICTED BY THE FILM CLOSE ENCOUNTERS OF THE THIRD KIND SO AS TO PREVENT ANY IMMEDIATE DISCLOSURE OF THE FACT THAT FIRST CONTACT ACTUALLY TOOK PLACE FIFTY YEARS AGO. THIS IS OnLY PART OF A CLASSICAL PSYCHOLOGICAL WARFARE PROGRAM DIRECTED AT THE AMERICAN PEOPLE BY THEIR OWN GOVERNMENT WITHOUT APPROPRIATE AUTHORITY. THERE IS ALSO A DON EFFORT IN PROGRESS TO FULLY IDENTIFY OTHER ALIEN ENTITIES WHICH HAVE HAD AN IMPACT ON THIS PLANET. SPECIFICALLY, THERE IS AN EBE SPECIES KNOWN AS THAT OF THE "BIRDMEN" WHICH IS BELIEVED TO BE SOMEHOW REPTILIAN RELATED. THERE IS ALSO AN ALIEN SPECIES WHICH HAS BEEN COLLOQUIALLY IDENTIFIED AS THE "BABIES" FOR THEIR SOMEWHAT NEONATAL APPEARANCE. (THESE MIGHT BE SOME SORT OF HYBRIDS AS WELL.)

THE INTERSPECIES RIVALRY CURRENTLY IN PROGRESS INVOLVS ELEMENTS OF WHAT CAN OnLY BE TERMED ESPIONAGE. IN OTHER WORDS, BOTH SIDES HAVE UTILIZED TACTICS AND STRATEGIES USUALLY ASSOCIATED WITH COVERT INTELLIGENCE AGENCIES. THE MOST STRIKING TACTIC EMPLOYED BY THE REPTILIANS IS THAT OF THE "FALSE FLAG" WHICH INVOLVS THE PORTRAYAL OF HOSTILE REPTILIAN

ACTIONS AS BEING THOSE OF MORE BENIGN SPECIES. THIS PARTICULAR TECHNIQUE HAS BEEN FOCUSED ON THE UNITED STATES GOVERNMENT ON SEVERAL OCCASIONS AND, IN ONE INSTANCE THAT WAS RECOUNTED TO ME, RESULTED IN THE KILLING OF NUMEROUS GREYS BY UNITED STATES MILITARY FORCES. THE DISSEMINATION OF DISINFORMETION IS ALSO A FAVORED TECHNIQUE. THE UTILIZATION OF MIND CONTROL TECHNIQUES BY BOTH SIDES MAKES THIS PARTICULAR ESPIONAGE GAME A DIFFICULT ONE TO COMPREHEND OR MASTER.

THE ABOVE IS EVERYTHING I CAN RECALL GIVEN THE FACT THAT I DID NOT KEEP WRITTEN NOTES FOR REASONS OF SECURITY. I FREELY ADMIT I KNOW THAT MUCH OF THIS DATA IS OF A CLASSIFIED NATURE. NEVERTHELESS, I CANNOT DETERMINE HOW MUCH OF THE ABOVE WAS FED TO ME AS DISINFORMEATION OR TO WHAT DEGREE I AM THE VICTIM OF MIND CONTROL. HOWEVER, I STRONGLY BELIEVE THAT I HAVE BEEN SOMEWHAT SUCCESSFUL AT MY INVESTIGATION AND THAT MUCH OF THE ABOVE DATA IS RELIABLE. FURTHERMORE, I CANNOT EVEN GUESS IF THIS INFORMETION, EVEN IF COMPLETELY ACCURATE, IS BEING DELIBERATELY LEAKED THROUGH ME OR IF IT IS ACTUALLY SOMETHING THE SHADOW GOVERNMENT WANTS SUPPRESSED. IN ANY CASE, IT IS MY CONCLUSION THAT MY SAFETY—AND THAT OF THOSE I CARE FOR—IS BEST PRESERVED BY THE...IMMEDIATE DISSEMINATION OF THIS INFORMETION. IN ANY CASE, THIS INFORMETION IS LARGELY UNVERIFIABLE AND CAN ALWAYS BE SUBJECT TO OFFICIAL PLAUSIBLE DENIAL. AS OF THIS POINT I WILL OnLY UNWILLINGLY HAVE ANYTHING TO DO WITH THE SUBJECT OF EXTRATERRESTRIALS OR WITH ANYONE ASSOCIATED WITH THIS FIELD.

AS THEY SAY, THE TRUTH IS OUT THERE, BUT CAN WE RECOGNIZE IT WHEN WE SEE IT?

*"We cannot tell you how badly we feel
about the bad ETs. They are like Nazis,
they have no regard for human life.
Humans are treated like mice,
bugs and laboratory animals.
They are doing this to be able to exist
on Earth with hybrid creations.*

*Ultimately, they will not succeed, as good
cannot be overcome by evil. These
despicable creatures will eventually leave
on their own. There are other ETs who
are not like them also living on your
planet. We are forbidden to intercede.*

*It is Earth karma that must be dealt
with by human consciousness. It is a
treacherous game Earth's team can win,
if the players are determined enough to
keep the ball in their court."*
– Zarg of Excelta

EPILOGUE TO THE REPORT:
OR
AM I STILL DEAF AND BLINDED TO THE TRUTH?

As I finished reading the report, I sat there sighing and shaking my head for what seemed like a long time. I did not physically want to put the report down, since it was apparent that was now going to be my last physical connection to Jordan. I knew it, I could sense that his statement that he didn't want to be connected with anyone connected with UFOs was sincere. I held the typewritten pages to my heart in hopes that I could feel Jordan's energy field. I hoped that I might sense in his vibrations the veracity of these powerfully alarming words.

If it was true that the aliens put us together to see how we responded to each other because of our connection, we provided a mighty good show for them and we never missed a cue. That is, *if* it was true! I kept remembering all of the times I said how much Jordie and I were bonded and how we were so alike! It came out of my own mouth, that was the only clear fact I knew for sure. Did I say it because it was a genetic fact or because I was under some kind of mind control? I sure hated to think that I was, but I also mentioned, myself, that might be a possibility. If a person is under mind control, would they *even think* they were? Or, are there levels of thought and actions controlled in a limited form, where you are able to evaluate your own situation? It is said that crazy people don't think they're crazy. What about mind controlled people, do they know they are?

As far as the rest of Jordan's report, there weren't many surprises. I couldn't help being more concerned about my own personal situation which still was not validated for me. I had a lot of questions unanswered by his words. I now doubted whether I ever would get these answers. Whoever was orchestrating the symphony prefers to play to the deaf.

141

I got up and walked over to my windows looking out over the azure blue of the ocean. I could see the tourists beginning to trickle back from the beach to their condos and hotel rooms. I smiled as I thought to myself, "All these people are probably thinking about right now is having a Margarita and getting into a hot shower, while I'm trying to figure out if my ovum were taken by aliens and whether I had sex with my own FBI agent son because of it! I guess Jordie was right about one thing, I am *not* the average woman!

I did have one more chance to speak with Jordie. He phoned me for the last time later that evening. After that call, I took a hot shower and had a Margarita.

THE BIG KISS-OFF

(The following is a transcript of the actual tape recorded conversation between Mia and Jordan with Jordan's full knowledge that all conversations were being recorded.)

JORDAN: I'm calling to say goodbye.

MIA: I'm upset at that. At least, if you had seen me in person, we could have communicated about this. The substance of your report is no surprise and I don't doubt what you said but the thing that really disturbs me is that you have no proof about me. I'd like to know how that was possible, how "they" got that information. I believe that was owed to me. It doesn't state anything specific about me. There is only a vague, parenthetical statement. How do I know that is me?

JORDAN: Why do you doubt that?

MIA: I have never heard of that technique related to the ovum. It doesn't feel right or make sense. Either I am wholly your mother or I'm not. If you believe I am your mother, how can you cut me out of your life? We are a part of each other.

JORDAN: Knowing who you are hasn't changed my feelings for you and that is wrong, it's perverted and immoral.

MIA: Why does that mean that we still can't see each other and be a part of each other's lives? You can't just will it or mentally erase me from your life. You can't tell your mind to not think of black, without thinking of black. You can't say, "I can't think of Mia," without thinking of Mia.

JORDAN: But you are not suppose to sexually desire your mother!

MIA: You are believing what "they" told you. Perhaps, we have another kind of connection and "they" are manipulating you. I am not going to plead the case for this, you are going to have to decide about this yourself

JORDAN: I have made my decision.

MIA: I guess you have and I feel badly because I'm bonded to you and I'll miss you terribly. You cannot fault yourself for sexual misconduct, for feeling the way you do. You didn't know!

JORDAN: I can't stop feeling that way.

MIA: You are eliminating me from your life without closure. It is not closure for me.

JORDAN: If I saw you, I wouldn't be able to control my feelings about you. Right up until I was told about you, I was ready to give up everything for you, just to be with you.

MIA: At what point did "they" tell you about this?

JORDAN: I don't know how to tell you how I was told.

MIA: When did you write the note in my book to meet you. You haven't had that book for months?

JORDAN: Remember when I told you that just in case we needed a means to get in touch, I had written something where you could find it?

MIA: No, I don't recall that. Just in case! You must have had an idea or feeling that we would need to do that. You can accept what they say because they say it. You and "they" have shown me nothing of proof. They could say that to keep us separated. I have nothing of substance. "They" haven't told me where this happened or how it happened.   How did your mother and I have this interchange and why were we selected?

JORDAN: Within my heart, I know it is true.

MIA: I know there is something that is true, but I don't know for sure that your truth is my truth. I wanted something more for proof.

JORDAN: I know that you know in your heart that it is true and if you do, it must change your feelings towards me.

MIA: I can't say that my feelings have changed but I can say that my appetite has not been satiated for proof about this, unless they have shown you proof, then perhaps you know. I'm just a mere footnote, a mini-fact in your report. I don't see anything that says it is me, other than what they told you. You had your Gulliver's Travels, your voyage to Never-Never land. I have nothing but frustration for over three months I feel used and abused in the process. I wish you could give me one fact, one piece of paper that has my name on it. As far as I am concerned, all they know is my social security number and how much taxes I pay! I am not satisfied. It is not enough. Your in the proof business and our mental link is strong. If I am your mother, than the apple doesn't fall far from the tree. I want proof!

JORDAN: We are tremendously alike. But on my end, this is perverted, it is wrong.

MIA: Why are you flagellating yourself? How would you have known about this? You couldn't have. You had a need to have someone in your life who loved you and that was me. I don't fault you for that. You weren't psychologically miscued. You just felt love and something special. What is wrong with that? How could you have interpreted it differently, at the time?

JORDAN: But now I know it and I should change.

MIA: Why didn't you meet me at the restaurant today?

JORDAN: Because the closer it got to the time, the more I realized that all I wanted to do was to hold you and be close to you.

MIA: "They" didn't tell you not to come?

JORDAN: No.

MIA: You can try to eliminate me from your mental vocabulary but it is not going to work. I will leave the door open for you to communicate with me, if you choose. I certainly don't feel guilty because I had no knowledge of this nor could I have conceived of this happening. I told you that when I was married, I went to fertility doctors, because I couldn't get pregnant. So, this would have been the last possible thing in my life that could have happened.

JORDAN: Yes, but you do have a son.

MIA: And I love him and don't want to lose him...and I also

have grandchildren, that means something to me.

JORDAN: Remember when I was at Nellis and I asked you what kind of love you have for me?

MIA: Yes, I knew anything I would have said would have been wrong! You could have gone either way. If I said that it was platonic, how would you have felt? No matter what it would have been, it would be the wrong response.

JORDAN: But knowing what you know now, have your feelings changed?

MIA: I can't explain how I feel, mostly due to mistrust I have had about you and I have had plenty of reasons for that. There has been nothing you have ever said to me factually that has ever born inspection. I don't know why you told me about Dr. Weeks, why you talked on the phone to a busy signal and all of the other crazy things you did. With all the mistrust, manipulation and mind games, I still have feelings for you. I wish that I could wash them away, like you want to. It would be a whole lot easier to do that than to say that this is someone I am connected to, who I will always be connected to! Did you lie to me about all of those things?

JORDAN: No. I think you will find that with what you know now, as time goes by, a lot more will be revealed. Not by me, certainly, but it will dove-tail in and make sense. And only in retrospect will you be able to say that no doubt or checking could reveal what "they" are capable of.

MIA: It sounds to me like you were the one who was under tremendous mind control and not me.

JORDAN: Yes, that is absolutely true. I have volunteered for things that I didn't know I volunteered for.

MIA: Does that include this gig with me in checking out your ex FBI brother, Mark?

JORDAN: That was completely engineered.

MIA: Was he a part of this.

JORDAN: Not to my knowledge.

146

MIA:	I hate to think that my life was totally manipulated by outside forces. I liked to think that I had domain over my very own existence.
JORDAN:	You are very naive.
MIA:	I see clearly now that I am. Wasn't that what we all have been trained to believe? I am a strong person but now it feels like I have been tilting at windmills for a greater part of my life.
JORDAN:	I don't know what it was like 50 years ago to be a human but it was qualitatively, completely different than it is now.
MIA:	I have had a taste of that. But I would like to know what it is about me that makes me so damned interesting to these intelligence agencies. In the last four years, I have been subjected to total immersion into the wacky world of surveillance by the unseen intelligence agency forces. Why have they focused in on me with such intensity, like no other abductee or UFO researcher I have ever heard about? You haven't told me that...

After that last question to Jordan, either he hung up or our conversation was terminated by unseen hands. We never did get to say the goodbye Jordan had intended. Jordan and all of the agencies knew that I recorded my phone calls as a protective device. Jordan prefaced every phone call with the statement that he didn't care if "they" were listening in. Despite this, it is obvious, that the question of who I am is still not to be answered, even by my putatively beloved son of the cosmos.

Jordan sounded very resolute in his determination to stem the tide of his feelings for me, by remaining incommunicado for the rest of his life. I will not put any obstacles in his path. With great surprise to myself, I found that the maternal, nurturing part of me that has been childless, easily welcomed the transition from lover to mother, even preferentially. However, I will never accept that I am truly part of his genetics until I get proof, *some* kind of proof!

147

With all of the mind control and manipulation, the hall of mirrors takes on this added dimension for me. Now when I step in front of the mirror, I see my face and it is faceless except for a big question mark over it. If I have to settle for the me I know now, I am quite happy. I wear my battle scars proudly, on the field of honor of my life, as I know it in this reality. And if one day I find out that Jordan is truly my son and his children are my grandchildren, I will be joyous, for it will not be a great tragedy but, instead, will be an incredible miracle. I say that with clear mind, controlled only by my heart.

"Reeducating your belief system helps demolish bad memories. You will get a lover more quickly if memory of past is free of emotional blame.
Lost loves make lasting baggage.
Don't make a lost love like a
taste of a mint in your mouth!
– Zarg of Excelta

A CONVERSATION WITH ZARG

"MANY LIFEFORMS PICK UP YOUR STRONG ENERGY, MIA. THEY CAN'T HELP IT, INCLUDING THE RIEGALIANS. THEY ARE A LOOSE GROUP NOT CLOSE TO THE AUTHORITY, LIKE MAVERICKS. WE CANNOT SANCTION THEM FOR DOING BAD THINGS. IT IS A UNIVERSAL COROLLARY. YOU DO NOT PERSONALLY HAVE ANYTHING TO FEAR FROM THEM, AS THEY RECOGNIZE YOUR STRENGTH."

— Zarg

MIA: *Is it really true that Jordan is a product of my genetic material by alien manipulation?*

ZARG: (pause) *YES.*

MIA: *Did it happen in 1962 like Jordan said?*

ZARG: *YES, BUT NOT AT THE TIME YOU THOUGHT IT HAPPENED. IT WAS AT ANOTHER TIME. THEY SHUT YOU OFF SO THAT YOU WOULD NOT REMEMBER THE EXPERIENCE. WE TOLD YOU THAT OTHER ETs KNOW ABOUT YOU AND ARE ATTRACTED TO YOUR VIBRATORY FREQUENCY AND KINETIC POWERS. THEY FELT YOU WERE A GOOD CANDIDATE. WE COULD NOT STOP THEM THEN.*

MIA: *Why didn't you tell me?*

ZARG: *WHY SHOULD WE, WHAT PURPOSE WOULD IT HAVE SERVED? YOU HAVE YOUR MISSION FOR EXCELTA, THIS DOES NOT FORWARD YOU, IF YOU ARE FEELING VICTIMIZED.*

MIA: *Do I have other children through this alien ovum harvesting?*

ZARG: *PROBABLY YOU DO. WE DON'T EXPECT THIS SITUATION TO OCCUR AGAIN, LIKE IT DID WITH JORDAN.*

MIA: *Will I ever see Jordan again?*

ZARG: *HE IS MARRIED TO HIS JOB AND HAS BEEN TOLD NOT TO CONTACT YOU. IN HIS OWN WAY, HE MONITORS YOU BUT NOT OFFICIALLY. YOU RECOGNIZED YOU HAD A CONNECTION BY YOUR FEELING BONDED TO HIM. DO NOT DWELL ON THIS. USE YOUR TIME ON EARTH FOR YOUR ADVANTAGE. TODAY IS THE ANSWER.*

A PHENOMENAL
ENGLISH ENCOUNTER

One month after my last contact with Jordan, a vestige of the relationship with him was felt in England. I embarked on a journey of research to England and Ireland with other UFO researchers and experiencers. We were to meet with our kindred spirits in these areas. It was to be a forum for mutual interest and debate over issues related to UFO phenomenon.

I was delighted and amazed at how much world wide experience was beginning to jibe with those we have in the United States. In prior years, much conjecture and scoffing was made by non American researchers about the nature of our reality. Now, it seemed that the paradigm shift in UFO experiences in non U.S. locales was moving towards a critical mass situation. World wide, we seemed to be sharing the same experiences and seeing the same manifestations of crafts, aliens, lights and ovum and sperm harvesting.

Our first meeting with researchers in a London hotel was with the Association for the Scientific Study of Anomalous Phenomena, (ASSAP). The meeting was to begin, as I took my seat in the front row. Next to me was a gentleman wearing glasses. He introduced himself as John. One of the ASSAP members conducted the meeting and said that the main speaker would be John Spencer. He acknowledged John Spencer, as the man who was seated next to me. I turned to John in amazement and glee. "Are *you* John Spencer?" I asked. "Yes," he replied. "I have a really good story to tell you. You have been a part of my life and you don't know it!" I whispered.

The meeting ensued with some consensus and much debating. John Spencer was quite logical and even-handed in his presentation. There was a member of ASSAP who took the position that all UFO experiences were a result of allergies to electromagnetic and other physical causation. He had written a book about that position and was pontificating on it. This caused a real brouhaha. Many of the local researchers were appalled at the apparent lack of good manners in this ASSAP member, since there were abductees and experiencers visiting.

Things were getting a little out of hand and felt a bit divisive. It was then that I felt compelled to do something that I had *not* planned to do, to maintain cohesiveness for our two groups. I did not plan to tell my tale of my relationship with Jordan, with its manifold results for me and bizarre implications for civilization. It was just too much detail and too outrageous, to tell at one sitting. But now there was dissension in the ranks and I felt compelled to tell about how John Spencer, unknowingly, had played a role in my life, in a small way.

I raised my hand to speak. I was called upon and rose to face the audience of Yanks and Brits seated there. I had never discussed this experience in public before, let alone before "foreigners and doubters." I began with a short summary of my relationship with Jordan and its mysterious outcome. "Now here is where the British and American connection comes into my life, " I stated. "One of the last times that I spoke with Jordan, he repeated over and over, a page number of an unknown book, telling me that it was important for me to find it. I told him that I didn't know what he was talking about. Jordan insisted that he had every confidence that I did know and with a quiet certitude, he hung up the phone and left me standing there in frustration. I walked around my apartment in terrible distress, feeling that if I didn't logically deduce what Jordan had cryptically referred to, I would miss something very important. I felt very frustrated. Soon, like a bolt out of the blue, I was drawn to my bookcase and reached into it, pulling out a book. It was John Spencer's book, *Phenomenon*."

I then explained how and when the message had been written in the book and the implications of it having been done. "John Spencer, the author, is the same man who is here, sitting next to me today! It never had occurred to me that I would ever meet the author of the book in which Jordan chose to leave his message."

There was a stunned silence from the audience and then applause. It seemed that I had done the right thing, to bring the two groups together. John rose out of his chair and hugged me. He said that he was very touched hearing me and felt that the synchronicity of our bond had meant much for the two groups, now brought together by a common sharing.

I wondered after the meeting, was it really an accident that I sat next to John Spencer? Was it just happenstance that he was even at that meeting? How many and how often could the synchronicities in my life be just cosmic surprises? Perhaps someone was looking into a special crystal ball of my future course and set this up? But which was planned first, the writing in the book or the meeting in London? Such possibilities no longer seemed to be too phenomenal, and such speculation no longer seemed to be too ridiculous!

*"Earthlings soon will have
the truth lie at your feet.
Be ready, be prepared,
man brings to Earth
the cost of greed.
Humans must be ready
for some hard times that
will affect their vanity.
Don't fear knowing the
destiny of Earth.
All has purpose."*
– Zarg of Excelta

A FEW COMMENTS FROM ZARG REGARDING CURRENT EVENTS ON EARTH:

ON RUSSIA:(1993)
"THE UNIVERSE IS SLOW BECAUSE RUSSIA IS MAKING GREAT DEMANDS ON IT LIKE A VORTEX. GORBACHEV WAS A LEADER OF LIGHT. THE WORLD MUST ACCEPT CHANGES AND ALLOW THEM TO BE MADE."

– Zarg

ON THE GULF WAR: (1991)
"FOOLS, FOOLS AGAIN! YOU HAVE ALLOWED THE FORCES OF EVIL TO AGAIN COME TOGETHER, TO PUT THE PEOPLE OF YOUR PLANET AGAINST EACH OTHER, TO GO BACK IN A TIME WHEN POORLY MOTIVATED INDIVIDUALS WHO ONLY CARE FOR THEMSELVES, LEAD YOU. WHEN WILL YOU LEARN?

WE CAN ONLY GUIDE. WE HAD HOPES THAT YOU WERE WALKING DOWN TO THE LIGHT. THE EXCYLES MUST LEAD AND SET AN EXAMPLE. YOUR GROUP IS ONLY A SMALL LIGHT IN A BIG FIRE. EXCYLES ARE ALL OVER THE WORLD. ADVENTURERS HAVE LEAD THE WORLD ASTRAY. PEOPLE ARE TAKING ADVANTAGE OF YOUR WEAKNESS. A LESSON TO BE LEARNED. ALL EXCYLES MUST SEE THAT REALITY.

YOUR WORLD IS BEING CONTROLLED BY THE ONES WHO CAN BENEFIT FROM CHAOS. EARTHLINGS ONLY CAN SEE THE NOSE ON THEIR FACES, ONLY LOOK TO FEATHER THEIR NEST ON THIS PLANET. CREATE ENERGY THAT WOULD DIRECT GOOD INSTEAD OF EVIL.

YOU DID NOT LISTEN! SEARCH YOUR HEARTS FOR THE TRUTH IN THE MATTER. EACH INDIVIDUAL HAS THE POWER! THE TRUTH IS WITHIN. NOT ONLY YOUR GROUP, BUT ALL OF THE EXCYLES ALL OVER THE WORLD MUST BELIEVE THAT THE ANSWER LIES WITHIN YOU, EACH ONE! BELIEVE IN YOURSELF AND THE POWER WILL PREVAIL. DO NOT ALLOW YOURSELVES TO BE DISTRACTED. THE MASTER WILL KNOW.

MASTER YOUR FATE! DECIDE YOUR OWN DESTINY! YOU ALL WAVERED! LEADERS DID NOT LEAD AND TEACHERS DID NOT TEACH! THE LIGHT IS WITHIN AND YOU MUST TRUST IN IT! IT IS NEVER TOO LATE. SEARCH OUT WITHIN YOUR HEART AND DISCOVER WHERE YOU FAILED. LET THIS MESSAGE BE REPEATED AGAIN AND AGAIN. YOU MUST TRUST THE LIGHT AND LIVE THIS LIFE IN IT."

– Zarg

ON THE HUBBELL TELESCOPE: (1995)

"THE PROBLEMS WITH THE HUBBELL WAS OUR TOUCH. THE NEW TELESCOPE WAS DAMAGED BY THE COMMITTEE, WHICH IS HIGHER THAN EXCELTA, BUT PART OF THE COMMITTEE IS EXCELTA. THE COMMITTEE IS NOT AN INTERGALACTIC COMMAND, IT IS A HIGHER LEVEL OF AUTHORITY. WE CHANGED THE TELESCOPE BY PUTTING THE LENS IN BACKWARDS BECAUSE THERE IS CERTAIN EVIDENCE WE DIDN'T WANT FOUND YET. THE TELESCOPE WAS COMING TOO CLOSE AND WOULD HAVE FOUND SOME INDICATIONS OF A HIGHER LIFE FORM. YOUR PLANET IS NOT READY YET. IT MUST CLEAN UP ITS ACT HOW CAN WE ACCEPT EARTHLINGS INTO THE HIGHER LIFE CYCLE WHEN THEY CANNOT GET AN UNDERSTANDING OF THE BEAUTY OF LIFE?"

– Zarg

CHAPTER TWO

THE EXCYLES

"Life to you flows like a river,
and time you see
has taken a new direction.
Like a bend in the river,
it flows on as a continuum
but the flow goes on
ever stronger."

– Zarg of Excelta

THE DREAM AND THE DOGTAGS

In retrospect, as I was growing up, there were several hints about what "truth" was to manifest, but I just wasn't spiritually mature enough to be a good guesser.

Most people float through their lives giving little or no credence to the "coincidences" and "intuitive flashes" they experience regularly. These are generally dismissed as "random accidents" or "imagination," and the act of considering them any further is often viewed as making a mountain out of a mole hill. Although I wasn't always quite so attuned as I am now to these phenomena, I have always had a lingering curiosity about certain memories that never seemed to go away. It wasn't until the unfoldment of recent events that I discovered just how meaningful these past events really were in finding out who I really am.

I was the first of three daughters born to blue collar working class parents. I lived in Chicago until I graduated from Roosevelt University, with the exception of one teenage year, when I attended the University of Tampa. Both of my parents were totally devoted to their children and did all they could for us. As a whole, unlike most children, I don't have any major complaints about my parents. I just couldn't wait until I was grown up enough to leave home and begin my life. Even as a very young child, I knew that my life in Chicago was not really what my life should be. I felt as if I was in a holding pattern, flying around seeking a destination, looking for a place to land. Where I eventually landed was not on anybody's navigational charts.

My recollections of a recurring dream had always perplexed me. The dream began when I was about four years old and lasted until I was six or seven. It was a dream that both spontaneously occurred and also one that I chose to have. I remember consciously choosing to have it because there was such a loving and nurturing feeling attached to this dream. I would snuggle down under the covers and close my eyes tightly and either visualize it or ask to have it. At this point in time, I am not quite sure which way it went. What I *am* sure of is the intensity of loving emotion that was attached to the experience.

The dream itself is quite clear to me even though many years have passed since I experienced it. I saw myself in what appeared to be a hospital room. At least, I felt as if I were in a hospital bed. The bed itself was banked up at the top and I was half sitting up. I felt quite cozy and happy. I certainly didn't feel sick and I always wondered why I was there since I *didn't* feel incapacitated in any way. There were many "people" surrounding my bed. I remember calling them "stars" but I don't know why. Perhaps it was because of the incandescence of their energy fields. I have tried hard to remember their faces but all I see is their robed and hooded forms. I felt the intensity of love like a white beam showering me with affection. Later on in life, when I looked at the great masters' renderings of the Madonna's beatific smile at the Christ child, it seemed reminiscent of the faceless gazes I felt from these "stars." They always said the same thing to me, "We love you and miss you and can't wait to have you back again with us!" If love is the operative word here, there is no doubt in my adult mind (remembering my child's mind) that Love, with a capital L, was what I felt. It was truly blissful to be encapsulated within it.

There was not much else to the dream except, upon waking and remembering the dream, I would think, "Gee, if I was sick and in the hospital, why didn't those people bring me some presents?" It was a source of constant perplexity. Nowadays, I'd give up gifts entirely to have that kind of conscious love beam aimed at me on demand, or even sporadically given at another's will. Maybe one day...!

As an adult, the dream was constantly bugging me. I tried very hard to figure it out. Psychologically, it seemed facile. Any first-year psych student could analyze it. Believe me, I tried! I thought that I must have felt very lonely since I was the first of three children, and in my childish way I felt neglected and conjured up this dream of being a "child movie star" who became ill in some way and was visited by other "movie stars" who knew and loved me. I thought that I would choose to have that dream or would mandate it when I felt that I needed some nurturing. It was a child's survival mechanism. For all intents and purposes, it certainly would satisfy the superficial examiner's hypothesis, but somehow it never satisfied me! As I became more attuned to meditation and spiritual concepts later in life, I took several dream workshops in an attempt for a perfect and complete analysis of the true meaning of this long-remembered dream. Little did I imagine how miraculous the actual answer would be, and how profoundly it would change my life.

Most children have periods of anger when they say that their parents are not truly their "real parents." I knew that I physically resembled my parents, but there was always a pervasive feeling of detachment that I never fully understood. I certainly loved my parents but I had a kind of haunting feeling that I really didn't belong to my family. I had a feeling of duality. I not only felt a part of my family, I also felt *apart* from my family. Logically, I knew that I was genetically connected but somehow emotionally, I felt as if I was a stranger in a strange land. There was a pervasive loneliness that did not seem to disappear. It would at times be in remission but would inevitably rear its ugly head and created a void that sought to be filled. By all accepted standards, I appeared to be "normal", but there were many times when I felt much less than that. As a teenager, I remember creating a poem with a melodramatic line, "I must have sinned a great sin to deserve such hell..." It was too bad that I didn't have a replacement" dream" to cull up to help cure my discontent. Thirty years later, my seemingly overly dramatic separateness would be understood and my childish guilt for feeling so would be assuaged.

Concurrently, with this feeling of displacement occurred my first experience of a super psychical nature. Even though I was an often morose teenager, I was still capable of normal teen age emotions. I had a boyfriend, Sheldon, who was in the Naval Air Force. We had a mutual friend who introduced us, after which I proceeded to correspond with "Shelly" and to date him when he was on leave. One summer weekend, he came home to Chicago. Chicago has terrific beaches, and as was the normal thing, we went to Montrose Beach with friends to party and enjoy the always too short Chicago summer. We all ran into the Lake Michigan surf and proceeded to have a hilarious time splashing water on each other, playing lake warrior games by climbing piggy back on the guys' shoulders and doing water battle. After about a half hour of frolicking, Shelly stood up and reached around the periphery of his neck and, in horror, exclaimed, "My dog tags are missing. I had them on when we came into the water and now they're gone. I must have them! I'm going on an important mission after my leave and if I don't have them, I can't go! We have to look for them." He proceeded to dive in the water, which was approximately 4-1/2 to 5 feet deep, and the rest of the gang followed. When the lifeguard saw the frenzied activity, he swam over to question us. When Shelly explained our search, the lifeguard shook his head doubtfully. "I've been watching you guys," he said. "You've been swimming and diving all over the area. It would be impossible to determine where those things fell off; it could be anywhere. And with the sandy bottom of this lake, the waves could have buried them. You'll never find them! It would take a miracle!" He then proceeded back to shore.

The lifeguard's negative predictions did not deter Shelly. He wanted those dog tags and encouraged us to keep looking. Here comes the weird part of this saga; I felt an urging to move in towards the shore at least 100 feet diagonally to the left of the group. I was in about three feet of water. I took my biggest toe on my right foot and burrowed it an inch or two into the sandy bottom and felt something metallic. I grabbed the object with the rest of the toes on that foot and lifted it out of the sand just above the water line. I couldn't believe my eyes! In my hand were Shelly's dog tags! I was truly astonished; I didn't know what to make of this, and how would I explain this to Shelly and my friends? No one would believe me. *I* didn't believe me! I was confused. Nobody was anywhere near me to have seen the event. In my smitten teenager's mind, I was mostly concerned that Shelly would be angry with me for what he could only perceive as a ruse. After a few seconds, I saw that there was no choice. Obviously, those damned things were important to him, so I raised my arms and proclaimed that I had found the treasure. Of course, nobody did believe me. The consensus was that I hid them in the top of my bathing suit, and there was no way I could convince them otherwise.

The rest of the day I pondered that experience. It, indeed, *was* a miracle but how did I do it? In those days, psychic phenomena was not a common topic of conversation. I had no one I could discuss this happening with. At the dinner table that night, I mentioned it to my family to garner some sort of a response. They seemed as perplexed as I, and we just went on with the rest of the meal.

I now realize that my psychic ability was not a one-shot deal. There is a great and powerful ability within me that I have inherited and nurtured not only in this lifetime, but also in past lifetimes. Although at the age of seventeen I felt like some kind of a freak for the experience, it has now become a cherished and most blessed facet of my being.

In retrospect, everything is in its most absolute and perfect order. I have been fortunate to have the vantage point of a great and most momentous paradigm from which to view the universe and my world within it. The dog tag experience seemed like an unexplained miracle at the time. It was the beginning of my conscious relationship to other miracles and synchronicities that expanded my awareness. There is a section from *A Course in Miracles* that exemplifies this:

> *"Miracles are both beginnings and endings, and they alter the temporal order. They are always affirmations of rebirth, which seem to go back, but really go forward. They undo the past in the present, and they release the future. A miracle is never lost. It may touch many people you have not even met, and produce undreamed of challenges in situations of which you are not even aware."*

"You do not choose your parents half way. They are assigned to you like a chore. Birth is only for karmic start. It is something you have to do to learn a way to make your own life. Parents must earn your respect as you must earn theirs. Giving respect to parents by virtue of their position only is a man made construct. Parent and child are mutual responsibilities, none easier than the other, both part of the pathways of life experience."

– Zarg of Excelta

PICNIC PICKUP

When I was a child, my family in Chicago had a very large cousin's club. It was basically the extended family from my father's side. His maternal grandmother had sisters and they had many children. Therefore there were a lot of cousins, A LOT OF COUSINS! No one was rich enough to have a home that could accommodate the multitudes, so the annual picnics were held in parks or forest preserves. There was enough space to allow the kids of all ages to run to keep the tykes out of their parents' hair.

There was one picnic that I have etched deeply into my mind. I was six and a half years old, and it seems a strange memory. My parents were *very* protective of their children; my father was definitely over protective! He watched us like a hawk. We couldn't leave the house on a cold Chicago day, unless we were entombed Mummy-like in every stitch of clothing we had. We were warned about crossing the streets, not taking candy from strangers, dilly dallying on our way home from school and about every blessed fear a parent can have about their children. My parents' utmost concern about us was never in doubt or in question. That is why this memory is so strange to me.

During the summer of this particular picnic, there was a most violent act against a little girl who was abducted and killed by an unknown assailant and left in a forest preserve in the Chicago area. It was a summer of pure terror for local parents and my parents were justifiably paranoid. We were warned over and over about not going with strangers and sticking to our friends and siblings like glue. I was also fearful. It was the constant topic of conversation. I was not the type of kid who purposefully bucked her parents. I had my moments, but generally not in front of a hundred relatives.

My memory is very strong. I remember leaving the group and walking away. Since it was a fine summer day, the park was filled with people, so no one really noticed my straying from the fold. I just blended into the crowd. I was drawn away for sometime. I was not fearful. I followed the road that wound through the park. It is then my memory is lost. The next thing I remember is a tall, slim and kind man bringing me back to my family's fold and saying good-bye to me with a smile. I know that some time had passed, since people were packing up to go and good-byes were being said. What seems totally out of synch with this scenario is that my always concerned parents didn't say anything to me about my absence. Nobody seemed to have noticed, although I knew that I had been gone and that I was brought back by a stranger. It was incongruous with my life as my parents' child and the fears of the girl's killing being rampant.

Now, that I know what I do about the modus operendi of the "abduction/contact" experience, it seems an understandable event in my young life. This picnic was part of the physical contact one-on-one's that I was having with my ETs. It coincided with the events at sleep hours, when I went to the "hospital." When I asked my ETs how they could take me away from my parents without their knowing, Zarg said that they made sure that they were both asleep or unaware. I am sure that the same thing occurred at the picnic, since the most baffling part was my parents' lack of concern about my absence. Perhaps there was some kind of time warp or stoppage of real time, to allow the ETs the freedom to deal with me on their terms for whatever they needed to do with me. That makes sense, because I now know that whatever has happened that seems really weird in my life, I now can attribute to my connection with my lost family and friends of Excelta and its moons.

THE HOT TAR ACCIDENT

Since I was about seven or eight years old, I have been quite conscious of my thumbs! That seems a kind of an unusual fixation for a person of any age, but my thumbs are, indeed, weird. My mother insists that I had perfect thumbs as a baby and as a toddler. There was no indication of anything imperfect about those digits. It does seem that Mom was correct. I checked out many photos of myself as a toddler and as a youngster and the thumbs are perfectly aligned, no sign of anything amiss, no unusual angularity, strange bending or freaky joints. By the time I was ten years old, these thumbs were bent totally like the necks of swans, with protruding lower thumb joints. It was quite apparent and often commented on. Kids can be cruel, even about strange finger configurations.

As bent as they were, I never suffered any pain from them. They looked like arthritic fingers of an old person. I had suffered no trauma or illness during my childhood that would have created such disfiguration. There were no hospitalizations or diseases that could have accounted for them. They always perplexed me. The older I got the more distorted they became. Now one of the thumb joints is total immobile. I wanted to know if there was a panacea to get them back in shape or if anyone knew how they got the way they are!

It soon became apparent that my perplexity was shared by the medical profession. The thumbs now are in an osteoarthritic state. I do *not* have systemic arthritis. To get to this point of deformity, one must have a systemic disease. That was what was so baffling for the arthritis specialists I sought. They couldn't figure it out. I was told by a top local specialist in arthritis that if I wanted an answer the only place to get it would be the Mayo Clinic in Rochester, Minnesota. If there was an answer, they would be able to find it.

I made my way to the Clinic and spent one full week being examined. The first day, I was examined by the head of the rheumatology department and spent two hours draped in a drafty hospital gown while various members of the department ran in and out, talking amongst themselves. When I finally could tolerate no more, I demanded to know what was going on! The simple answer to me was that they were checking the medical text books and the computers to see if they could find a case similar to mine. Even with a world-wide computer network, they could not find one case! I truly am unique. They were so upset that they said they assumed I must have had a case of juvenile rheumatic fever which is a total absurdity since I would have known and so would my mother if I was that sick as a child. Many doctors seem to have egos that won't allow them to say that they don't know sometimes.

The final diagnosis of the Mayo Clinic left me with as much of a mystery as I came with. I still did not know for sure the causation of the deformities. The hand surgeon insisted that I must be in pain and I insisted that I was not. He did not seem to believe me. There was no way to cure the condition. The only thing suggested for aesthetic purposes only was surgery by fusion of the joints. I, also, found that to be inappropriate for me, at least at this time. After I left Mayo Clinic, I went to visit my sister in the Chicago area. When she questioned me about the results of the medical evaluation, I remember I just shrugged my shoulders and said they really couldn't tell me anything definitive. After a short pause, I said, "I guess I must have been taken on board a space ship and had an operation done on me when I was young." I remember that she looked at me quizzically, "What did you say?" I just laughed and walked away. I really had no idea why I said that! I didn't even know anything about aliens or space ships. In retrospect, that statement in itself proved that the layers of unconsciousness were beginning to peal away and I already knew the answers I sought at Mayo Clinic. I just didn't *know* that I knew.

Early in my conversations with Zarg, I asked him if he knew why my thumbs were the way they were. He said that it was the result of a "hot tar accident aboard a craft that occurred during some heated discussion." When I asked for further illumination on the subject, Zarg dismissed it as being unimportant. He said that it was just related to the physical and was not as important as the spiritual and that there was not much that could be done about the problem. When I protested it was important to me, he suggested some palliative, homeopathic remedies that did not help much. I had the feeling that he preferred not to discuss this. I also had the intuition that perhaps it was *not* Zarg's Exceltan ship that we were talking about but some other space aliens'. Even though Zarg readily admitted that other extraterrestrials know about and monitor me, he does not like to discuss this issue, saying that is for my safety. This is one area of my life that I cannot get any straight answers from either humans or non-humans!

In 1994, I attended a UFO conference where Linda Moulton Howe, the doyenne of animal mutilation theorists, was speaking. In an update on her research, she reported that the latest mutilations showed a "tarry substance" around the excisions. When I heard that I was startled. Was it possible that the new report was related to the mystery of my thumbs? Was it really true that there was a medical experiment "accident" and my digits were reminders of that? No matter, with the swan's neck like arch of the joints providing a natural hitch hiker's thumb, perhaps it provides the perfect mode for me to attract attention to hitch a ride out to the galaxy. It's worth a shot!

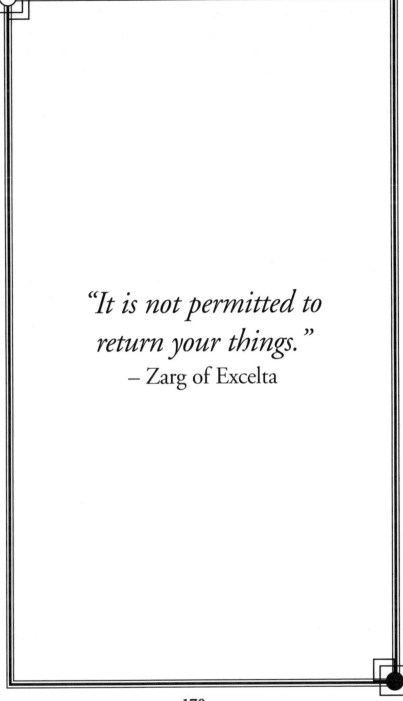

"It is not permitted to return your things."
– Zarg of Excelta

WHO DO THE VOODOO?

It felt unreal to be in Africa! I had no preconceived ideas as to what to expect. But here I was in Kenya and I loved it already. We were to start out in Nairobi and spend two weeks on safari. We were each allowed to take only one suitcase with us in the van and any additional ones we arrived with would be left in a locked storage room in our hotel in Nairobi. Off we went on our very special adventure.

From Nairobi, we headed on to Tsavo National Park and our first night was spent at Kilaguni Lodge. It was spectacular to see our first wild animals and we were thrilled. My roommate was Penny, a chum from Ft. Lauderdale. We were resting in our room, I was unpacking my suitcase and Penny was sitting on her bed smoking a cigarette and watching me. We were rehashing the first day's events. I took out my new pair of stone washed designer Capri pants and put it on my bed. Next to it, I laid an extra large heavy cotton shoulder bag with a picture of a kangaroo and the word "Australia" on it. I intended to use and well fill that bag with purchases while on safari. We continued to chat while I repacked.

When I turned back to get my pants and the bag to put into my suitcase, I was astonished to find that they were not there. I pulled the bed linen apart and looked under the bed but I couldn't find those two items. Penny asked what I was looking for and when I told her, we were both completely baffled! Where could they have gone? This was a subject that continued to haunt me. They were two of my most favored possessions and they virtually disappeared before our very eyes! I could not figure this out, and every night while I was on safari I looked inside my suitcase; I still could not believe they were gone.

On the next to the last day of our safari trek, we were staying at the Lake Nakuru lodge. The rainy season was just beginning, and the rain was falling heavily. We always maintained a prompt schedule, and that morning we were slated to leave for the Masai Mara and a balloon ride over the Mara. Since it was quite a distance in the rain to breakfast, I had to make an instantaneous decision whether I was going to go into my suitcase for my umbrella. I decided to do it thinking that since I had dragged it all the way from Florida, I might as well use it! I had purchased it specially for the trip and it still had the price tag on it. I opened it up and ran all the way to the dining room under a massive down-pour. When I got into the dining room, I slipped it under my chair and proceeded to chow down.

After finishing breakfast, I looked at my watch and saw that I had better move on to the van, as it was almost time for our group to leave. I didn't see any others from my group so I hurried off. En route to the van, I realized that I had left my new red umbrella under my chair in the dining-room. I stopped in mid track to decide what to do. The group was all in the van except me and our safari guide was motioning me on.

Well, I shrugged to myself, I guess that umbrella will be another piece of me that I leave in Kenya besides my pants and my Australia bag! I moved on to the waiting van.

Unfortunately, the downpour had made the mud roads leading to the Masai Mara impassable. The bridges were also washed out. We were all quite disappointed, but it would provide us with an extra day in Nairobi. The van turned around and we made our way back to Nairobi.

When we got to the hotel, we were told that our extra suitcases that were left in the storage room would be returned to us. Penny and I went to our room to rest and soon the extra suitcases appeared. I opened my locked suitcase and to my absolute astonishment found the same red umbrella with the price tag on it that I had left that morning at Lake Navasha! I was so shocked that I had to sit down on my bed. I turned to tell Penny about this but I couldn't get the words out. How could I tell her about something I didn't believe myself! I sought a rational explanation but could not find it. I knew that I had taken just one new red umbrella and had left it many miles away in the bush country. I thought that it must be some kind of voodoo powers of the Dark Continent that were beginning to affect me or that some kind of spell was on me. First the disappearances, and now the reappearance of an item left miles away magically manifesting in a locked suitcase. I didn't know what to think.

Within four months, I would have my answers to the many riddles of my life so far. Talk about magical...the truth of what actually happened was beyond any feat Houdini could have dreamed of.

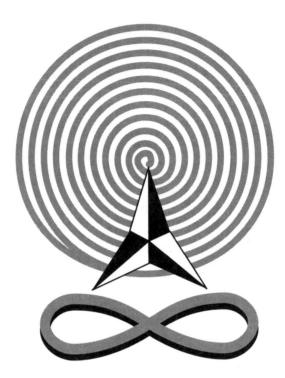

"Days of Righteousness and Greatness"

THEY CALL US THE EXCYLES

Little could I have guessed that sultry Sunday in August, 1988, it would be the day that my life would change forever. My friend Vanessa and I were having a pleasant post-church luncheon. We always looked forward to this time together to gossip and share what was going on in our lives. I treasured my time with Vanessa, as she was not only knowledgeable from a spiritual and metaphysical point of view, but she also had a wide gamut of real life experience to draw from. We were trying to decide on something to do post lunch to brighten up our mid summer doldrums, when Vanessa remarked, "Do you still have that Ouija board that your friend gave you? Why don't we see what the board has to offer us? I haven't done that for such a long time, it will be fun!"

Earlier that year, my friend, Jacque had asked me if I knew anything about Ouija boards. My knowledge of Ouija boards was limited and my interest was nil. I equated the boards with spirits who were most often capricious, and with humans who used the board for fun and games. I gave the Ouija board absolutely no credence.

Jacque told me that she had a Ouija board. She said that she had been very anxious to use it and had begged her husband, Marty, to use it with her. Since it takes two people to do the Ouija, he begrudgingly agreed. They sat at the board for several minutes, with both of their finger tips poised on the indicator. Marty was getting impatient since the indicator failed to move even a centimeter. He started to protest about the, silly, damn board, when spontaneously all of the pictures on the walls in their small apartment fell off the walls with loud crashes! Jacque and her husband were both startled and then became frightened. Marty jumped up out of his chair and away from the board. " Get this goddamned thing out of the house! I never want to see it again!" he exclaimed. He went off to inspect the damage of the crashed pictures. He was convinced that spirits from the board caused the incident.

The Ouija board had found its adopted home in one of my closets since Jacque was forbidden to keep it. It had remained there, unused, ever since the few board sessions I had begrudgingly agreed to with Jacque.

My reply to Vanessa's suggestion was less than enthusiastic. "Yes, I still have it but I haven't worked the board since Jacque went back to St.Louis several months ago. I think I packed it away in the closet somewhere. Do you really think that the Ouija board has any validity? I always did a little prayer before we used it because I didn't want to bring in any bad spirits. When Jacque and I used it, it never told me anything. It just seemed focused in on Jacque with this spirit called Darren that kept on insisting on Jacque's getting pregnant so he could be born and she could be his mother! I have heard so many negative things about the use of that thing."

"Now, Mia," Vanessa replied, "if you use the white light of protection and are surrounded by the Intelligence of the Creator, it is a pretty sure bet, that if you bring in anything on the dark side, you will know it and won't be affected. It is our choice as to what energies we choose to deal with. You know that you are an individualized part of God Mind, so you will attract kindred energies. If it doesn't feel right, all you have to do is stop using the board."

I felt more encouraged and comforted by Vanessa's words. We went back to my apartment and I found the Ouija board where I had put it, on the top shelf in my closet. We set our chairs close together near the windows overlooking the ocean. The board was placed between us on our laps. Despite Vanessa's assurances, I didn't want to take any chances with negative energies so I lit a white candle and burned some incense, as I had heard that they purify your environment. It then occurred to me that I should say an affirmation before we did anything.

Seemingly out of nowhere the words came to me and in my mind, they felt just right. We closed our eyes and I said,

"THE PEACE, THE POISE AND THE POWER OF GOD IS WITH US NOW. THERE IS NO SITUATION, THERE IS NO CONDITION, THERE IS NO MEMORY OF THE PAST, NO IDEA IN THE PRESENT OR VISION OF THE FUTURE, THAT CAN SEPARATE US FROM OUR INEXTRICABLE TIE WITH THE INFINITE DIVINE, CREATIVE INTELLIGENCE THAT IS GOD IN OUR LIFE. WE GATHER HERE TODAY TO REUNITE OUR CONNECTION WITH THOSE WHO ACKNOWLEDGE AND LOVE US ON THE UNSEEN. WE ASK THAT THEY USE US ONLY AS CHANNELS FOR GOOD. WE KNOW THAT WHATEVER INFORMATION WE GET, IT BE FOR THE HIGHEST AND BEST OF ALL CONCERNED. WE SURROUND OURSELVES WITH THE WHITE LIGHT OF PROTECTION, KNOWING THAT THIS INFORMATION WILL NOT BE INJURIOUS TO OUR MENTAL, SPIRITUAL, EMOTIONAL OR PHYSICAL BODIES. IT IS THEN WITH GREAT HUMILITY AND HUMBLE GRATITUDE, WE ASK THOSE ON THE UNSEEN LEVELS TO MAKE THEMSELVES KNOWN TO US TODAY. AND WE AFFIRM ONLY INFORMATION THAT MOVES US FORWARD. AND SO IT IS, SO BE IT, AMEN."

After I said the affirmation, I felt more secure and ready for whatever action the board would provide. Neophyte as I was, I had no real expectations. I certainly was not interested in getting pregnant, so I didn't think the little spirit Darren would ever come back again, I was definitely a lost cause in his direction! Surprisingly, it didn't take long for the planchette, (the Ouija word for indicator), to start moving. It began to circle the board and then slowly found its way to each letter and then number. When it started the move around the numbers for the second time, I said to Vanessa, "Are you doing that? There is so much energy, you must be moving this thing!"

"No, Mia, I am not doing a thing except holding on to this planchette for dear life!" she retorted. "Be quiet and let's see what it is saying!"

"GREETINGS, GREAT THINGS HAPPEN NOW, MIA AND VANESSA"

"Wow, whoever you are, you certainly seem to know your way around this board. You are so quick. Who are you?" I asked.

"I AM ZARG, FROM EXCELTA."

"Hi Zarg. That is a neat name. Where is Excelta, I don't recall ever hearing of that place before?"

"EXCELTA IS NOT ON EARTH, IT IS ON ANOTHER LEVEL OF SPACE AND CONSCIOUSNESS."

When I read these words, I was very surprised. It had never occurred to me that the Ouija board could bring in anything other than disembodied spirits, those who were still Earthbound and hadn't found there way into the light yet and those who had already made their transition to another plane of existence. Vanessa was smiling broadly. "Well Zarg, we are so happy to talk to you. I am sure that you have some interesting things to tell us. How did you happen to come to speak with us today?" she asked with much interest.

"MIA'S TOTO ATTRACTED US THROUGH HER VIBRATORY FREQUENCY."

"What?" I laughed, "My toto! Now that sounds kind of sexy, what is that?"

"TOTO IS THE EXCELTA WORD FOR SOMEONE WHO HAS THE ABILITY TO SEND OUT AND RECEIVE VIBRATORY FREQUENCIES. MIA YOU CAN DO THAT.

"Gee, that is something I didn't know about myself. Is that like extrasensory perception?"

"THAT IS PART OF IT BUT NOT ALL. YOU WILL LEARN MORE ABOUT THAT LATER. WE HAVE MORE IMPORTANT THINGS TO TALK ABOUT."

"Really," I responded, "like what?"

"IT IS NOW TIME THAT YOU KNOW WHO YOU REALLY ARE. YOU ARE AN EARTHLING BUT YOU HAVE NOT ALWAYS BEEN. YOU AND VANESSA ARE FROM EXCELTA.. YOU BOTH ARE SPIRITUALLY MATURE ENOUGH TO KNOW THE TRUTH ABOUT WHO YOU ARE. YOU ARE EXCYLES FROM THE HOUSEHOLD OF KING DUGUD. YOU CAME HERE WITH HIM A LONG TIME AGO."

"Who is King Dugud, do we know who he is?" Vanessa asked.

"YOU KNOW HIM NOW AS TROY PALMER. YOU MUST BRING HIM HERE TO SPEAK WITH ME. THE PROCESS OF UNFOLDMENT MUST BEGIN NOW SO THAT ALL EXCYLES MAY PREPARE THEIR WAY HOME TO EXCELTA."

"Troy Palmer from our church?" Vanessa and I exclaimed in unison. We were speechless for the moment. We both were friends with Troy. I had met Troy and Dan O'Hara on a church trip to Hong Kong. The three of us were sitting together. I didn't know either of them until then. By the end of our plane trip westward, we became close buddies and shared conversation as if we were long lost friends. Then just at the moment that I was thinking about how I met Troy, Zarg began to move the planchette again.

"DAN O'HARA IS ALSO FROM THE HOUSEHOLD OF KING DUGUD, THAT IS WHY YOU ALL WERE DRAWN

179

TO EACH OTHER FOR THE TRIP."

"Yes, I was just thinking about that, how did you know?"

"WE ARE COMMUNICATING TELEPATHICALLY. I CAN HONE IN ON YOUR TOTO, JUST AS YOU ARE HONING IN ON MINE. DON'T WORRY ABOUT MY KNOWING WHAT YOU'RE THINKING. WE ON EXCELTA HAVE BEEN TUNING IN ON BOTH OF YOU FOR A LONG TIME AND KNOW YOU WELL. THERE IS NOTHING YOU CAN THINK OF THAT WOULD SHOCK OR SURPRISE US. WE LOVE YOU UNCONDITIONALLY. YOU MUST BRING TROY TO US SOON. HE IS ALSO READY TO KNOW. THERE ARE OTHERS WHO YOU KNOW THAT ARE FROM EXCELTA AND ITS MOONS. YOUR TOTO, MIA, HAS DRAWN THEM ALL INTO YOUR LIFE. YOU ARE ALL FAMILY. THE EXCELTA FAMILY HAS STRONG BONDS. YOU WILL KNOW WHO MANY ARE SOON. SOME ARE NOT IN YOUR SPHERE RIGHT NOW BUT WILL BE. SOME ARE NOT AS PREPARED AS YOU BOTH ARE. SOME WILL ACCEPT THE TRUTH ABOUT THEIR HERITAGE AND OTHERS WILL NOT BE READY. YOUR JOB IS TO TELL THEM WHO THEY ARE. YOU ARE PIVOTAL, MIA. YOU MUST DO THIS NOW. THE TIME HAS COME FOR ALL TO KNOW WHO THEY ARE IN THE UNIVERSAL SCHEME."

Talk about being prepared, I was *not* prepared for this message. Although, I don't know if anyone can really be prepared for this kind of information from a Ouija board. The thing that really amazed me was the mention of Troy's name. Troy was my good friend and was an excellent metaphysician, but maybe I had underestimated him. At least from an Excelta viewpoint, I had! Again as before, Zarg was reading my thoughts.
"TROY IS A VERY POWERFUL MAN, MORE THAN HE KNOWS. HIS POSITION AS KING WAS NOT

MONARCHAL. IT WAS A TITLE OF HIGH SPIRITUAL AND PHILOSOPHICAL ATTAINMENT.. WE CALL IT TRUSTEE BUT IT IS ALSO LIKE A KING OR FIGUREHEAD OF A SPIRITUAL HIERARCHY."

"Perhaps, it is like our Pope?" I equated.

"CLOSE BUT NOT EXACTLY. YOUR POPE HAS ONE POINT OF VIEW THAT NOT ALL EARTHLINGS ADHERE TO. KING DUGUD WAS TRUSTEE TO ALL OF EXCELTA AND STILL LIVES IN THE HEARTS OF THE PEOPLE. HE WAS EDUCATED IN THE HIGHER WISDOMS AND PERFECTED REASON."

"If he is so well loved and important, why did he and his household come to Earth?" Vanessa quizzed.

"EXCELTA IS EVOLVING LIKE EARTH. IT IS ON A VERY HIGH SPIRITUAL PLANE, BUT LONG AGO THERE WAS A PROBLEM BECAUSE OF AN UPRISING OF THE GREEDY SEEKING POWER. IT WAS LIKE A COUP D'ETAT ON EARTH. THOSE THAT LOVED HIM SOUGHT TO PROTECT HIM BY SENDING HIM AND HIS HOUSEHOLD AWAY. AT FIRST, HE WENT TO A PLACE LIKE A MONASTERY TO HIDE UNTIL IT WAS SAFE. BUT THEN SOME THOUGHT THAT THEY HAD KILLED HIM BUT THEY MADE A MISTAKE, IT WAS SOMEONE ELSE. IT WAS DECIDED THAT THE KING AND HIS HOUSEHOLD MUST LEAVE EXCELTA UNTIL IT WAS SAFE TO BRING HIM BACK, WHEN THERE WAS NO LONGER A THREAT."

"Why was Earth chosen?

"EARTH WAS CHOSEN FOR TROY'S EXODUS BECAUSE THE LIKELIHOOD OF HIS EVER BEING DISCOVERED WAS NIL. EARTH WAS OF NO IMPORTANCE FOR THOSE

AT THE EXCELTA LEVEL AND WAS VERY FAR AWAY FOR SHIPS OF THAT TIME. NOW IT IS DIFFERENT. MORE ARE AWARE AND CARE ABOUT EARTH BECAUSE OF THE DIFFICULTIES THERE. YOU HAVE A VERY STRANGE PLANET WITH MANY BAD PEOPLE. YOUR PLANET ATTRACTS MANY E.T.s, BECAUSE OF THE WIDE RANGE OF EMOTIONS AND EXPERIENCES THAT ARE NOT AVAILABLE ELSEWHERE. YOU ALSO MUST HELP EARTH WITH YOUR CONSCIOUSNESS. SHOULD YOU CHOOSE TO USE YOUR SPIRITUAL RESOURCES, EXCYLES ARE VERY POWERFUL, SPIRITUALLY."

"I am kind of confused here, this is starting to sound like Star Wars or Star Trek," I said. "Do you know those shows?"

"YES, THEY ARE FUNNY BUT NOT REALLY THE WAY IT IS. THEY SPEAK WITH THE BIG VOICE AND NOT THE SMALL VOICE. I SEE THEM THROUGH YOU. SO DO THE OTHER BROTHERS AND SISTERS HERE ON EXCELTA."

"What does that mean big voice?"

"MOST OTHERS NOT FROM EARTH SPEAK TELEPATHICALLY THROUGH THEIR MINDS, NOT VOICES."

I wanted to get down to some of the more mundane questions like what they looked like and where they lived in the universe.

"WE ARE NOT PHYSICAL BEINGS IN THE EARTH SENSE. WE ARE LIGHT BEINGS AND HAVE BODIES OF PURE ENERGY. AS NEEDED, AT A MOMENT OF IMPORTANCE WE HAVE THE ABILITY TO MANIFEST IN THE PHYSICAL, BUT IT IS NOT OUR FIRST CHOICE. IT TAKES TREMENDOUS POWERS AND ABILITIES TO DO THAT. WE CAN DO MOST ALL OF OUR WORK IN LIGHT BODIES."

"Are you like angels or spirit guides?"

"NO THEY ARE TOTALLY DIFFERENT PARTS OF GOD'S PLAN. THERE ARE HELP MATES AT EVERY LEVEL OF WHAT YOU CALL LIFE AND DEATH. ALL OF GOD'S CREATION IS LIKE A JIGSAW PUZZLE MOVING IN SPACE. EACH PART IS INTERRELATED AND INTEGRAL. IF ONE PIECE IS MISSING YOU DON'T HAVE THE TOTAL PICTURE. THERE ARE HELPERS AND GUIDANCE IN EACH PIECE OF THE PUZZLE; SOME OF WHAT YOU THINK IS GUIDANCE OUTSIDE OF YOURSELF IS JUST YOUR HIGHER SELF OUTPICTURING IN THE REALITY THAT YOU CREATED AS YOUR LIFE. REST EASY ABOUT THIS, SOON ALL WILL BE REVEALED AND YOU WILL HAVE A SHINING LIGHT OF UNDERSTANDING."

"You mentioned that his enemies thought that Troy, King Dugud was killed. If he was a light being, which I believe means you have eternal life, how could he be killed?"

"I WAS USING YOUR WAY OF SPEAKING. ENERGY IS NEVER LOST, IT CAN ONLY BE TRANSFORMED. I WAS DESCRIBING A COSMIC DEATH THAT DESTROYS MOLECULES AND WHERE ATOMS BURST PHYSICALLY. IN THIS CASE, HIGH FREQUENCY VIBRATORY WEAPONS WERE USED TO TRANSFORM FROM ONE LIGHT FREQUENCY TO ANOTHER. IT IS A DIFFICULT CONCEPT TO CONVEY AT YOUR LEVEL OF SCIENTIFIC LITERACY."

"But how is it that there is such negativity at your high level?"

"THERE IS WHAT YOU CALL YIN AND YANG IN ALL THINGS. ALL ENERGY MUST HAVE ITS POLAR OPPOSITES, IT IS WHY SATAN BECAME THE FALLEN ANGEL OF GOD. SATAN IS A CONCEPT OF THAT POLARITY."

"Where is Excelta, is it in our galaxy?"

"NO, EXCELTA IS AN INTERDIMENSIONAL PLACE AND A LEVEL OF CONSCIOUSNESS. IT EXISTS AT A REALM WELL BEYOND THE EXISTENCE OF THE PHYSICAL EARTH AND ITS CONSCIOUSNESS. IN ATTAINMENT OF CONSCIOUSNESS, THERE ARE 22 LEVELS OF DIMENSION THAT ARE DIVIDED INTO THREE PARTS, SEVEN, SEVEN AND EIGHT. EACH HAVE PLATEAUS WITHIN EACH DIMENSION. EXCELTA IS AT THE 22ND LEVEL, THE TOP OF THE LAST EIGHT."

"But, there still seems to be conflict and strife on Excelta? That doesn't seem to be elevated, consciousness-wise."

"WE HAVE SINCE MOVED TO A HIGHER REALM AS WILL EARTH AT YOUR NEXT MILLENNIUM. WHEN KING DUGUD WENT TO EARTH, EXCELTA WAS AT THE FOURTH LEVEL OF THE LAST EIGHTH. WE HAVE EVOLVED SINCE THEN. THAT IS WHY THOUSANDS ARE PREPARING THE WAY FOR THE EXCYLES AND THE RETURN OF KING DUGUD. WE CANNOT COMPLETELY MOVE ON UNTIL THAT TRANSPIRES. MIA MUST HELP PREPARE THE EXCYLES

"What happens once you reach the 22nd level?"

"THAT IS IT."

"Where is God?"

"OVER ALL THE LEVELS."

"Do you ever see God?"

"IN THE MIRROR, EVER PRESENT IN MAN AND LIGHT BEINGS."

"What is it like at the 22nd level?"

"IN A SENSE, IT IS LIKE HOW YOU WOULD FEEL IN UTOPIA."

"What, besides getting the Excyles back, do you do at that level?"

"WE ARE UNIVERSAL WATCHDOGS. WE SEE AND LISTEN FOR THE ADVANCEMENTS OF ALL COSMIC PLANETS, THAT IS WHY WE TRAVEL SO MUCH. WE NEVER INTERFERE IN THE PROCESSES OF EACH PLANET, WE JUST MONITOR THE PROGRESS AND DEVELOPMENT."

"Do you need a space ship to travel in if you are light beings?"

"YES, WE TRAVEL USING GROUP MIND AS OUR POWER SOURCE. FEMALE ENERGIES GUIDE US AND ARE MOST OFTEN THE PILOTS AND NAVIGATORS, AS THEY ARE GENERALLY MORE INTUITIVELY FOCUSED THAN MALES. WE DO HAVE A VEHICLE OF LIGHT THAT CAN BE TRANSFORMED TO A PHYSICAL CRAFT. ALL IS CREATED OUT OF THE POWER OF INTENTION. WE LEARN HOW TO DO IT TOGETHER. ON EARTH, YOU CALL IT A HIVE MENTALITY AND NEGATE THE PROCESS. WE DO NOT LOSE THE POWER OF OUR OWN INDIVIDUATION, WHEN WE BAND TOGETHER MENTALLY. IT IS DONE WITH THE FEELING OF BENEFICIALITY FOR THE WHOLE GROUP. LOVE THROUGH MIND PROVIDES THE DESIRED RESULT. IT IS A SIMPLE CONCEPT BUT DIFFICULT TO EMBODY. THAT IS THE SECRET AND THE GOAL OF THE GALACTIC MIND. IF ALL OF THE UNIVERSES AND THE MULTI-UNIVERSES WOULD USE THIS TECHNIQUE, EVERYONE WOULD BE AT THE 22ND LEVEL AND ALL LIFE WOULD BE UTOPIAN. UNFORTUNATELY THE

LOVE OF GREED, POWER AND SELF INTEREST SUPERCEDES THE LOVE OF GOD WITHIN ALL BEINGS AT THE LOWER LEVELS AND EVEN AT SOME OF THE HIGHER LEVELS. THE EARTH IS NOT THE ONLY PLACE IN THE UNIVERSE THAT IS HAVING TROUBLE WITH THIS CONCEPT. THAT IS WHY LIMITED GUIDANCE IS SOMETIMES PROVIDED, ALTHOUGH NOT BY US, TO HELP THINGS MOVE EXPEDITIOUSLY."

"How do the soul and the spirit participate in this evolvement?"

"THE SOUL LEVEL BECOMES THE STANDARDIZATION OF MAN'S DIVINE PRINCIPLE. EXPERIENCES ARE RECORDED IN SPIRITUAL AWARENESS. THE GREATER THE AWARENESS, THE MORE THE SOUL EVOLVES."

"Are we able to move from one dimension to another after each lifetime?"

"IN THE FIRST SEVEN LEVELS, CRUDE TRAVEL IS POSSIBLE AND INTERSPERSED. IN THE SECOND SEVEN, SOULS MOVE THROUGH CONSCIOUS CHOICE, AND IN THE LAST EIGHT, SPIRITUAL LIVES ARE IN TRUE BALANCE WITH THE SOUL AND THE LEVEL CHOSEN IS APPROPRIATE FOR SPIRITUAL ATTAINMENTS. THAT IS WHY KING DUGUD AND HIS HOUSEHOLD MOVED DOWN LEVELS ONLY OUT OF THE UTMOST NECESSITY. IT IS HIGHLY UNUSUAL."

"Where is humanity now?"

"MAN IS NEARING THE LAST THIRD OF THE FIRST SEVEN. WITHIN EACH SEVEN THERE ARE STAIR STEP LEVELS."

"What you have said is contrary to what I have read and been taught here on Earth. Due to my egalitarian sensibilities, I have a hard time accepting that I am from the highest place, I don't feel that spiritual or like an avatar."

"IT IS LIKE LOOKING AT THE MOUNTAIN, INSIDE OF THE MOUNTAIN, EXPERIENCING THE MOUNTAIN, BEING AT ONE WITH THE POWER OF THE MOUNTAIN AND ULTIMATELY SEEING THROUGH THE MOUNTAIN. MAN IS NOW LOOKING AT THE MOUNTAIN. THE FIRST SEVEN GOES THROUGH ALL OF THESE STAGES. YOU ARE JUST A REFLECTION OF WHERE THE CONSCIOUSNESS OF YOUR PLANET IS. THAT IS WHY I HAVE CONTACTED YOU TO TELL YOU THAT YOU ARE MORE, MUCH MORE THAN THE FIRST SEVEN. YOUR TRUE HERITAGE, AS AN EXCYLE AND YOUR EVOLVEMENT ON EARTH EXPRESSES WHO YOU REALLY ARE."

"Well, I wasn't raised to feel smug and superior, it isn't my way."

"THAT IS YOUR EARTH TRAINING, IT ISN'T YOUR COSMIC PARENTAGE. GET OVER IT. HUMILITY IS OVERRATED ON YOUR PLANET. IT IS A WAY HUMANS CREATED TO KEEP THE MASSES ENSLAVED. THE ONLY HUMILITY SHOULD BE IN REVERENCE TO THE GOD WITHIN AND WITHOUT. YOU SHOULD NEVER USE HUMILITY TO KEEP YOU IN LACK AND LIMITATION. HUMILITY IS NEVER AN EXCUSE TO KEEP YOU FROM BEING THE BEST YOU CAN BE ON YOUR EVOLUTIONARY PATH. HUMANS HAVE BEEN TRAINED TO LOOK AT GOD IN THE SAME WAY THEY REGARD A MOUNTAIN, THEY ARE STILL AT THE LEVEL OF THE FIRST SEVEN. MOST ARE STILL JUST LOOKING AT GOD, NOT PASSING THROUGH AND EXPERIENCING GOD."

Vanessa and I looked at each other totally fascinated at the tenor and quality of the information we were receiving. It came quickly and with much energy, never sluggish, like it was sometimes with other entities. It certainly seemed very spiritually based on God and not any dark forces. Again without my asking, Zarg continued.

"OF COURSE, OUR BEINGNESS IS ENMESHED WITH GOD. LOOK AROUND YOU AND TRY TO DENY THAT GOD EXISTS! ALL LIFE IS A PRAYER."

"The way that you speak to us and the fact that you are a light being seems almost as if you are a god to us."

"NO, WE ARE EQUALS ON DIFFERENT PLANES."

"Are you closer to God than we are."

"NO, THE SAME. WE CAN WATCH OVER YOU AND HELP KEEP YOU SAFE BECAUSE YOU SHARE OUR PROTECTIVE LIGHT. WE ARE ALL FROM THE SAME ENERGY. EXCELTA BROTHERS AND SISTERS CARE CLOSELY FOR EACH OTHER. WE CANNOT SEPARATE OURSELVES FROM OUR PAST. YOU ARE PART OF US. WE HAVE CREATED OUR LIVES TOGETHER. IT IS OUR OWN DECISION."

"Did those on the Excelta level come from the same planet?"

"NO, EXCELTA IS ONLY A RESTING PLACE, A STATE OF CONSCIOUSNESS ABOVE IT ALL. WE ALL CAME FROM MANY PLACES AND LIVES. ON THIS HIGH LEVEL, THERE IS A SPECIAL SPECTRUM OF LIFE FORMS."

"I still don't understand why at that level there was so much dissension between beings that King Dugud and his household had to leave Excelta."

"NOT ALL SPIRITS ARE HAPPY LIGHT BEINGS THAT HAVE REACHED THEIR FULL POTENTIAL. EVEN NOW, SOME ARE STILL ATTACHED TO THE IDEA OF PHYSICALITY. MANY ARE SOMETIMES JEALOUS OF THE EARTH BOUND AS THEY MAY NOT HAVE PROSPERED AS A PHYSICAL BEING. YOUR METAPHYSICAL IDEAS ON EARTH ARE NOT ALWAYS CORRECT. ONE DOES NOT HAVE TO BE PERFECT TO BE A LIGHT BEING. IT IS STILL A LEVEL OF SOUL WORK AND EXPANSION. ONE MAY REACH PERFECTION AT A DIFFERENT LEVEL. WE STILL HAVE OUR POWER STRUGGLES, AND WE ARE STILL PERFECTING OURSELVES. IT IS TOO BAD THAT WE ARE STILL STRUGGLING, BUT IT IS PART OF THE PROCESSING OF CAUSE AND EFFECT. THAT IS WHY GETTING THE EXCYLES BACK IS SO IMPORTANT. IT IS A COMPLETION, KARMICLY. YOU SHOULD ENJOY EARTH NOW. WE MUST ALLOW YOU YOUR OWN DECISIONS. LATER THERE WILL BE MUCH TO BRING US TOGETHER."

"Can you understand my doubtful questions?"

"IT IS GOOD. YOU ONLY WILL GROW IF YOU ASK QUESTIONS. THE ANSWERS ARE THERE AND WAITING TO BE DISCOVERED."

"How long have you been watching us?"

"MANY EARTH LIVES. YOU WILL HELP OTHERS TO UNDERSTAND NOW. YOU DID NOT LISTEN BEFORE. THE EARTH IS MOVING TOWARDS AN UNHAPPY TIME. WE WILL ALWAYS BE THERE TO PROTECT YOU."

"Is this a time for punishment from the Creator for Earth?"

"THERE WILL NOT BE A PUNISHMENT, BUT EARTH

WILL GO ANOTHER WAY. THERE WILL BE ANOTHER LEVEL OF CONSCIOUSNESS ON EARTH. EARTH IS MOVING ON ITS DESTINY. THAT IS WHY THE EXCYLES ARE COMING TOGETHER. MIA IS THE MAGNET THAT BRINGS EXCYLES TOGETHER FROM VARIOUS PARTS OF THE EARTH LEVELS. YOU MUST BE CLOSE TO THE TRUTH. EXCELTA WILL GUIDE SO YOU WILL KNOW THE WAY."

"Can the Excyles make a difference together through the changes?"

"YOU WILL GET TO KNOW OTHER GROUPS BUT MOST IMPORTANT IS HELPING INDIVIDUALLY THROUGH EXPANSION OF YOUR UNDERSTANDING. THE UNITY OF EXCYLES WILL REINFORCE YOUR ENERGY SO THAT EACH WILL HAVE THE ANSWERS THEY SEEK. THERE ARE EXCYLE GROUPS FORMING ALL OVER THE WORLD. TODAY IS A DAY TO REJOICE AND BE HAPPY AS TO WHO YOU ARE AND AFFIRM YOUR HERITAGE. TAKE CONTROL, SET A DIRECTION AND THEN FOLLOW IT. THE SPIRITUAL FATHER OVERSEES YOUR GROWTH."

"What about Excyles who fall by the wayside and ones who won't listen?"

"THEY MUST SEARCH ALONE. THERE ARE OTHERS WHO HAVE CHOSEN NOT TO HEAR THE MESSAGE. YOUR TIME ON EARTH IS TO LISTEN AND TO LEARN. THE EXCYLES ARE LOST SHEEP WHO MUST NOT WANDER TOO FAR FROM THE TRUTH. TOGETHER YOU CAN OVERCOME WHAT HAS BEEN PUT IN FRONT OF YOU TO DETER YOUR UNDERSTANDING. TOGETHER YOU MUST DEVELOP AND KEEP A STRONG LIFE. YOU CAN MAKE A DIFFERENCE, SINCE A PEBBLE IN A POND CREATES RIPPLES THAT MOVE OUT. MIA, YOUR MISSION HERE IS TO INFORM AND

ACKNOWLEDGE OUR FRIENDS, THE EXCYLES. YOU MUST LEAD ON THE PATH OF UNDERSTANDING. WE ARE COMING CLOSE TO THE TIME WHEN WE ALL WILL BE TOGETHER. YOU MUST KEEP THE TRUE FAITH AND EVENTUALLY YOU WILL WRITE A BOOK FOR EARTHLINGS. VANESSA AND THE KING WILL HELP YOU."

"Well, that is a lot to put on me. I am just a beginner in this. I am not even sure I believe all of this. It is heavy duty fascinating metaphysics. I am flattered that you think I am something special but I am not so sure I can bring people to rally around a Ouija board."

"YOU CHOSE THE OUIJA BOARD THROUGH YOUR CONSCIOUSNESS BECAUSE IT WAS EASIER. IT WAS A WAY THAT YOUR SPIRIT MANIFESTED TO POINT YOU IN THE RIGHT DIRECTION. YOU CO-CREATED THAT REALITY WITH JACQUE. IF THE SITUATION WOULD HAVE CHANGED AND JACQUE WOULDN'T HAVE GIVEN YOU THE BOARD, WE WOULD HAVE FOUND ANOTHER WAY TO CONTACT YOU. YOU REALLY DON'T NEED THE BOARD TO COMMUNICATE WITH US. YOU CAN CHANNEL US TELEPATHICALLY, BUT YOU MUST LEARN TO HAVE FAITH IN YOUR OWN ABILITIES. THAT WILL COME IN TIME. THE BOARD WILL BE THE BEST MODALITY TO BRING THE GROUP TOGETHER IN THE BEGINNING. HUMANS NEED SOMETHING TANGIBLE OR EXPERIENTIAL TO MAKE THE UNREAL SEEM LOGICAL. THAT IS WHY WE TELEPORTED YOUR CLOTHING AND UMBRELLA IN KENYA, WHILE YOU WERE ON SAFARI THIS YEAR. WE KNOW HOW YOUR MIND WORKS. LIKE OTHER HUMANS, YOU NEED PROOF BEFORE YOU HAVE FAITH. MIA, WE ARE REAL IN YOUR TERMS. WE ARE NOT FIGMENTS OF YOUR IMAGINATION OR VANESSA'S. YOU MUST HAVE FAITH

191

IN US. YOU MUST OPEN YOUR MIND AND REACH FOR THE STARS!"

Again my mouth fell open, I was much more than surprised! The mystery of my disappearing items and my reappearing umbrella in Africa was part of the master plan of Zarg and Excelta! I could also assume the pictures falling off the walls in Jacque's house was also a part of getting this board to me. Even my cynicism was beginning to erode because the synchronicity of events towards this end result was overwhelming. How could I not be in awe at the orchestration of these events for the purpose of this contact! I was beginning to get goosebumps just at the idea that beings on a level of spiritual attainment, of whom I was not even aware of consciously, went to so much trouble to seek me out!

"Do you know what goosebumps are Zarg? You are giving me them with all of this proof?"

"YES, MIA, I KNOW WHAT GOOSEBUMPS ARE, THEY ARE LOVE MESSAGES, AND YOU GOT YOURS STRAIGHT FROM EXCELTA."

"To say that I am overwhelmed at the ideas you have given to us today, is an underestimate of my feelings. You have given us a lot to digest. I can't wait to tell Troy who he is. I bet he'll be surprised. One thing I know is that if I was to have made this all up, I would have made myself and Vanessa the Queens of Excelta."

"MIA, YOU WERE QUEEN UNSURLA ON EXCELTA BEFORE TROY WAS KING. EXCELTA WAS A SMALL NATION OF SOULS THEN. TROY WAS KING WHEN EXCELTA WAS POWERFUL. HE MADE IT SO AND IT HAS STAYED THAT WAY SINCE. THE MEMORY OF KING DUGUD HAS KEPT EXCELTA STRONG. GETTING THE KING HOME IS A CLOSELY GUARDED SECRET. WE WILL DIRECT YOU FROM NOW ON TO MAKE SURE YOU ARE

SAFE. MIA, YOU ARE THE HIVE AND THE EXCYLES ARE THE BEES."

"I would like to believe this but I am feeling some fear about accepting all that you have said unconditionally. I am afraid that it is like giving my power away to an unknown entity and I am frightened to do that. It seems reasonable to be frightened."

"ALL WILL BE WELL IF YOU USE YOUR INTELLIGENCE INSTEAD OF YOUR EMOTIONS. FEAR IS AN EMOTIONAL RESPONSE THAT IS AN ERROR IN ADDRESSING ISSUES. IT IS DISEMPOWERING, SO YOU MUST NOT MISTAKE EMOTIONS FOR INTELLIGENCE IN THE DECISION MAKING PROCESSES IN YOUR LIFE. IF YOU HAVE DOUBTS, MEDITATE. GO DEEPLY WITHIN WHERE TRUTH DWELLS AND FEAR IS RELEASED. YOUR ANSWERS RESIDE THERE. IT IS THERE THAT COMMUNICATION WITH ALL OF THE UNANSWERABLES OF LIFE ARE FACILITATED. WE HAVE CONTACTED YOU NOW BECAUSE WE BELIEVE THAT YOU ARE CAPABLE OF ACCESSING TRUTH BY TRIUMPHING OVER YOUR FEARS. YOU HAVE ATTAINED A CERTAIN LEVEL OF MASTERY AND GROWTH THAT WILL GAIN YOU ACCESS TO THESE TRUTHS AGAIN, AS YOU HAVE IN AGES PAST. WE WILL NOT AND CANNOT FORCE YOU TO ACCEPT WHO YOU REALLY ARE, YOU MUST ACKNOWLEDGE THIS YOURSELF. WE ARE HOPEFUL THAT YOU WILL AND CAN, BUT WE WILL LOVE YOU NONETHELESS, IF YOU DO NOT. THINK ABOUT ALL THAT WE HAVE SHARED WITH YOU AND VANESSA TODAY. PLEASE BRING KING DUGUD, TROY, TO SPEAK WITH US THE NEXT TIME. JUST REMEMBER THAT WE LOVE YOU BOTH AND THAT YOU HAVE FAMILY ON EXCELTA WHO HAVE NEVER FORGOTTEN YOU AND ARE HAPPILY EXPECTING YOU HOME SOON. WE EMBRACE YOU BOTH IN OUR MINDS AND HEARTS. GOOD BYE."

"Well wasn't that something," Vanessa said.

"Something aint the word for it," I joked. "I am going to call Troy right now."

I phoned him and luckily he was home. "Hi Troy, guess what...!

"We want you all here.
Make us an offer we can't refuse!"
– Zarg

"Excyles have all been aboard craft while young
and in other planes or realities while in the
sleep state. You forget the trip but you are
energized. We ask your permission before we
bring you on the ship. You are not in a
conscious state."
– Zarg

"Days of Righteousness and Greatness"
– Motto of King Dugud

King Dugud's Crest

Spiral Symbol   = *Deep into inner soul*

Three Pointed Star = *Righteousness, Integrity, Strength and Structure*

Infinity Symbol   = *Greatness and Justice. Justice is infinity because Truth prevails ultimately.*

"Excelta speaks through you, you must truly carry the message of love and peace to your planet to all, Excyles and Earthlings alike."
– Zarg of Excelta

EXCELTA ON EARTH

My revelations about Troy's cosmic royal birthright didn't seem to affect him as being incongruous. I must say that I was surprised! I thought he would make a joke about that, since he is very adept at turning most situations into humorous ones. He listened as I gave a general accounting of what transpired with the first Ouija board contact. I tried not to make it too serious of a reality, in case Troy thought it was fodder for his jokester cannon. He listened to every word I uttered without questioning or commenting.

When I finished, Troy said to me, "Wow, that is incredible. I always had a feeling that I had some connection with E.T.s. As you were telling me your board experiences, I started to think about my childhood in Cleveland. I remember that when I was about four or five, I had this little kid sized chair and I would take it out into the backyard just as the sun was setting and I would sit out there until it was dark. I would stare into the sky as the stars were beginning to twinkle and I clearly recall thinking that I belonged out there, that my home was in the stars! I did this as often as I could, even though my brothers would make fun of me. I couldn't and I didn't want to stop. I just wanted to be on my little chair and look at the night sky. And when I gazed up at the stars, a great wave of loneliness, like homesickness, came over me. As I got older, I wasn't so compulsive about doing that but I never forgot the feeling of homesickness. Now, I guess, I know why! So when are we going to speak to Zarg?"

For the second time that day, I got goose bumps! The quiet intensity of Troy's words really got to me. I could picture him on his little chair and then I remembered my own childhood feelings of being out of place in my reality, never being totally happy, always a little bit apart from and not a part of my world.

Could this all be true? "Well, babe, why not," I thought to myself, "Zarg told me to keep an open mind." In my life I had seen that synchronicity led me down important paths of self discovery that became integral in my personal evolution. "Troy it is up to you, whenever you want to give it a go, I'll be ready for another round with Zarg."

As we set up a time and date, I got this very strong image in my mind of Snoopy doing his happy dance. I then heard the words mentally, "Mia, that's us on Excelta now that you and the King are going to speak to us. We love you!" As I saw Snoopy in my mind and heard those words in my mind's ear, I knew in my heart that this *was* all true. I understood that this vision inside my head was what my brothers and sisters on Excelta called the "small voice" or telepathy, and that I did have that ability to be a radio antennae to the unseen realms. It also seemed to be at a feeling level. I could actually feel the joy of the Exceltans as they monitored my conversation with Troy. They knew, just as Troy and I now did, that as Zarg had said, GREAT THINGS HAPPEN NOW!

Troy, Vanessa and I met together for Troy's first Earthly reunion with Zarg. As soon as we began the transmission Zarg asked for Troy to come on the board. When Troy's fingertips touched the planchette, the thing practically levitated and became like a whirling Dervish pointing to each letter faster than I could comprehend the letters' combined meanings. The energy was palpable, and Zarg seemed to be in ecstacy speaking to Troy.

"THE KING HAS A GIFT OF PURE LOVE. HE IS THE EXALTED ONE. HE GIVES LOVE TO MANKIND. KING DUGUD HAS GRACE FROM GOD. YOUR KARMA ON EARTH IS NEARLY COMPLETE AND YOU WILL RETURN TO ALL THOSE WHO HAVE NEVER FORGOTTEN YOU AND HAVE BEEN WAITING FOR YOU IN YOUR TRUE HOME. YOU AND MIA ARE LIGHT BEINGS IN A UNIVERSE THAT CONTAINS MANY BILLIONS OF LIGHT BEINGS, BUT NOT AS BRILLIANT AS BOTH OF YOU. YOU ARE EVOLVING ON GREATER PLANES OF EXISTENCE

AND ARE AMONGST THE ONLY FEW THAT EVER ACHIEVED IT ON EARTH. WE ARE JOYOUS IN OUR REUNIFICATION WITH THE ENERGY OF OUR LOST COMRADES, BROTHERS AND SISTERS AND KINDRED SPIRITS. NEVER AGAIN FEAR THAT YOU ARE ALONE, FOR OUR LOVE FOR YOU BOTH WILL PROVIDE YOU WITH CLOAKS OF GREAT POWER AND PROTECTION TO SAFEGUARD YOU FROM HARM AND EVIL. TROY WITH MIA'S HELP, YOU WILL GATHER YOUR ROYAL HOUSEHOLD ONCE AGAIN. MANY EXCYLE SOULS ARE IN YOUR ENERGY FIELDS NOW. YOU WILL RECOGNIZE THEM, AS THEY ARE ATTRACTED TO YOUR LIGHT. MOST HAVE NOT RAISED CHILDREN THIS LIFETIME AND HAVE FORMED NO TRUE EARTH FAMILY UNITS. THEY HAVE BEEN GIVEN THE TIME AND OPPORTUNITY FOR SPIRITUAL PATH SEEKING, WHICH IS OFTEN DELAYED OR DETERRED BY THE RESPONSIBILITIES OF THE MUNDANE WORLD. THERE ARE OTHER EXCYLES FROM THE MOONS OF EXCELTA THAT HAVE FOUND FAMILIAL RELATIONSHIPS APPROPRIATE FOR THEIR SOUL EXPANSION. THEY ARE NO LESS IMPORTANT AND ALSO WILL BE BROUGHT INTO THE FOLD. INTRODUCE THEM TO THEIR TRUE COSMIC TIES AND HOPEFULLY, THEY WILL ASCEND TO THE STARS WITH YOU. DO NOT BE TOO ATTACHED TO THE RESULTS OF THEIR ACCEPTANCE OR REJECTION OF THEIR HERITAGE. YOU CAN OFFER THEM A FULL PLATE BUT IT IS THEIR CHOICE TO SATISFY THEIR HUNGER. STRENGTH COMES FROM BONDING. THE STRONG GETS WEAK WHEN NOT NURTURED. EXCYLES MUST BOND TOGETHER. TROY AND MIA ARE THE GLUE THAT KEEPS THE BONDS STRONG. YOU MUST LISTEN TO THE VOICE OF EXCELTA. LIFE ESSENCE COMES THROUGH EVERY DAY WITH EACH SUNRISE. OUR LOVE AND OUR ENERGY FLOWS LIKE RAIN DROPS INTO YOUR SOULS. ALL EXCYLES MUST

SHARE THEIR LOVE AND LIGHT. THE TIME IS COMING CLOSE TO BE TOGETHER. LOOK AHEAD, DON'T LOOK BACK, BUILDING FOR THE FUTURE WHILE DRAWING STRENGTH FROM EACH OTHER AS YOU GET CLOSER."

Troy could barely whisper his responses to Zarg, he was so touched by his words. Somewhere deep within his being there was a resonant chord struck, a harmonic vibration that entered and thrilled his spirit and the spirit of that child within who had gazed so longingly at the stars.

Troy Palmer was just 40 at the time of his first board contact. He had not been king in the world of Earthly realities but had lived a life of ups and downs just like average mortals. He is a tremendously creative, a real "can-do" person, with hands that seem to know how to fix, assemble and create anything. He had decided to be a furniture designer and was in the throes of dealing with a new business, at that time. Zarg did not mince his words of wisdom when suggesting how we deal with our day to day realities in the mundane world. He was especially kind, supportive and loving to Troy. Even though Troy was a brunette, he was Zarg's fair haired, Golden Boy. Not that he was unkind to the rest of us Excyles, but there was no attempt to mask his partiality. Whenever the other female Excyles and I complained that Zarg was just a male chauvinist, Zarg responded that he was not but that Troy was who he was and he must be treated accordingly.

Zarg was not the only Exceltan to speak with us. It seemed that many, from all spiritual and job levels wanted the opportunity. From General Wapet, the commander of the project to spiritual counselors to navigators, male and female energies, all wanted to extend their greetings to the Excyles and especially the King. Often, we would find out that these individuals had a connection to us in the past on Excelta and we could feel the delight they had in delivering the information to us. What was truly amazing to me was that they each had their own personality, even when delivering words of sweetness or wisdom or just plain fun. I was heartened that even at higher spiritual levels, even in scenarios where communal focus is paramount, individual persona and humor was not lost. Zarg was especially fast with the quip, even using the most esoteric American colloquialisms. When I asked how that was accomplished, Zarg said it was done effortlessly through mind to mind link. If it was in our mental language center, it became his. That interrogatory launched his next lesson.

"YOU MUST KEEP YOURSELVES PURE AND ON THE WAY TO EXCELLENCE. LISTEN TO THE VOICE THAT SPEAKS TO YOU FROM YOUR BROTHERS FROM SPACE. CARRY ON AND GO FORWARD AND BE GOOD TO EACH OTHER. YOU MUST BE STRONG OF DIRECTION, MEND YOUR BODIES AND STRENGTHEN YOUR MINDS."

"What do you mean by keeping ourselves pure?"

"TO BE PURE IS TO HAVE A THOUGHT PATTERN IN AN AWARE STATE OF COMPLETE AND WHOLE CONSCIOUSNESS IN A CURRENT FRAME OF TIME. YOUR BODY, MIND AND SOUL BECOME EQUAL ENERGIES. WHEN YOU ARE THAT, YOU ARE IN PERFECT BALANCE."

"Now that sounds like quite an accomplishment. I don't know if that is possible on the Earth plane. I was hoping you were just talking about purity as sexual celibacy. That has been a lot easier to accomplish for me, I can't speak for the rest of the Excyles."

"WE ARE NOT ASKING YOU NOT TO LOVE. WE ARE ASKING YOU TO LOVE BIG, TO STRIVE TO ATTAIN A DIRECTION FOR THE LOVE OF MANKIND."

"WITHIN BOTH OF YOU, YOU ARE THINKING THAT MY REQUESTS OF YOU ARE TOO UNATAINABLE AT YOUR LEVEL OF SPIRITUAL COMPETENTCY. YOU ARE VIEWING YOURSELVES AS TROY AND MIA OF EARTH, SEEING YOURSELVES WITH 20th CENTURY EARTH EYES. WE ON EXCELTA SEE YOU FROM OUR PERSPECTIVE OF YOUR DEVELOPMENT AT THE TIME WHEN YOU LEFT US. WE KNOW WHO YOU ARE, YOU HAVE FORGOTTEN YOURSELVES AND THE EONS OF TIME THAT HAVE CREATED YOUR BEINGNESS. PICTURE YOURSELVES AND YOUR SOULS' DEVELOPMENT IN HUGE MOMENTS OF TIME, THOUSANDS OR MILLIONS OF EARTH YEARS FROM NOW. YOU WOULD HAVE ADVANCED GREATLY. FROM YOUR PERSPECTIVES YOU WOULD BE LOOKING AT YOURSELVES AND VIEWING YOURSELVES AS SPIRITUALLY GOD-LIKE BEINGS! WHAT YOU WERE WHEN YOU LEFT IS STILL A PART OF YOU NOW, IT IS JUST DORMANT. TO AWAKEN THE SLEEPING GIANTS WITHIN YOU IS NOT AN EASY TASK, SINCE YOU HAVE HAD THE CONDITIONING AND THE ASSIMILATION ON A PLANET THAT KNOWS ONLY THE WORD GOD AND IS JUST IN INFANCY AT UNDERSTANDING WHAT GOD REALLY IS. ALL OF THE EXCYLES DERIVED FROM A SPIRITUAL PLATEAU MUCH ADVANCED OF EARTHLINGS. YOUR JOB AND ESPECIALLY MIA'S IS TO PROVIDE THE CATALYST TO POKE THE QUIESCENT CHRYSALIS, AWAKEN THE BUTTERFLIES FROM THEIR COCOONS. EXCELTA SPEAKS THROUGH YOU, YOU MUST TRULY CARRY THE MESSAGE OF LOVE AND PEACE TO YOUR PLANET TO ALL, EXCYLES AND EARTHLINGS ALIKE. WITHIN THE FABRIC OF THEIR BEING, EXCYLES WILL RECOGNIZE AND HEED THE CALL OF THEIR PAST BIRTHRIGHT AND THEIR IDENTITY AND CONNECTIONS TO SPIRIT. AS ALL WALK TOWARDS THE LIGHT OF UNDERSTANDING, OTHERS WILL JOIN IN AND THERE

WILL BE A REVOLUTION IN CONSCIOUSNESS AND A HARMONIOUS OUTCOME FOR ALL PEOPLES OF THE EARTH. THEN EARTHLINGS WILL STOP HAVING IMPERIALISTIC WARS OVER RELIGIOUS DOGMA AND IDEAS OF SECTARIAN RELIGIOUS SUPREMECY IN FAVOR OF LOVING ALL OF MANKIND. THEN PROGRESS WILL HAVE BEGUN AND IN UNITY ALL WILL WALK TOWARDS GOD AND START TO UNDERSTAND GOD AS MORE THAN JUST A SUPREME BEING BUT AS A SUPREME ACTION. YOU WILL ACTIVATE A CALL TO ARMS, AND WILL ENGAGE THE MIGHTIEST ARMY OF LIGHT EVER SEEN ON EARTH."

"There is no life experience that is of such great value that it exceeds each being's connection with oneness with the divinity."
– Og of Excelta

TELEPATHIC COMMUNICATION:
A COSMIC CASINO-GRAM

Two weeks after the first Ouija board contact, I went to Las Vegas, Nevada. I joined two friends from Tennessee, Linda and Sandy, at a famous hotel on the Strip and had planned to have a great time. These friends knew nothing about the contact of my space family from Excelta. I wasn't even sure how to tell them or if I should but destiny provided an avenue for discussion.

My ex husband was living in Las Vegas and we met for lunch at Caesar's Palace Hotel. We spent a pleasant time together, when he had to leave on a business appointment. I decided to play black jack at Caesar's. As I was looking for a seat at a table, I got a "beeper," a hunch to move my action to the Sands Hotel. I wasn't familiar with the Sands but I just felt intuitively that I should go there. I walked to the Sands and as soon as I entered, I had another mental beeper to sit down in the lounge and have a soft drink first. I found a table, ordered a drink and then noticed the Keno pads. Instead of using them to bet on numbers, I took out my pen and started writing. I had a real compulsion to do that. I felt a bit out of place and strange doing that in the midst of the sounds of clacking chips and slot machines' bells but it seems that I had no choice, the need was so strong! This was my first opportunity to receive conscious telepathic communication and to channel it to automatic handwriting. The message came so quickly and fluidly, that it was effortless. I was still bowled over with the fact that there were those beings on the unseen level who even knew who I was, let alone acknowledged me. So this kind of communication was a major revelation! The following are the words that I received in the most unlikely of venues, a bar in a Las Vegas hotel.

OG SAYS:

DON'T LOOSE FAITH IN YOURSELF! FOLLOW YOUR INSTINCTS, EVEN THOUGH THEY LOOK FAULTY, THEY ARE NOT. YOU STILL WILL FIND TRUTH BY

THE END OF THIS STAY. KNOW I AM WITH YOU FOREVER. LIFE IS A JOURNEY DIVINELY INSPIRED AND YOU HAVE ALWAYS BEEN CLOSE TO THE INSPIRATION. DAY AFTER DAY, LIFE NEVER FALTERS, FOR YOUR LIFE'S COURSE WILL FOLLOW ITS OWN CHART. YOU MUST BELIEVE THAT AND KEEP IT IN MIND. YOU ARE A BEACON THAT ATTRACTS AND PROMOTES DIRECTION FOR OTHERS WHO NEED YOUR LIGHT. TRUST YOUR INSTINCTS, THEY ARE GOD-DIRECTED. LIFE IS A SIEVE THROUGH WHICH EXPERIENCES FLOW, THE FLOW IS ALWAYS BACK TOWARDS THE SOURCE OF ALL. SOME EXPERIENCES FLOW EASILY, OTHERS ARE TOO FULL BODIED TO FIT IN THE HOLES WHICH IMPEDE PROGRESS. THESE SUBSTANCES MUST BE WATERED DOWN WITH KNOWLEDGE AND WILL EVENTUALLY FIND THEIR DESTINATION. THERE IS NO LIFE EXPERIENCE THAT IS OF SUCH GREAT VALUE THAT IT EXCEEDS EACH BEING'S CONNECTION WITH ONENESS WITH THE DIVINITY.

I was so astounded that I was writing this, I asked myself how this could be done. I once again received a mental reply.

WE SEE YOU MENTALLY, AS YOU CAN MENTALIZE US. WE HEAR YOUR THOUGHTS WHEN WE WANT TO TUNE IN TO THEM. ALL OF YOUR THOUGHTS ARE GOOD TO US, AS WE SEE GROWTH IN THEM, SO DO NOT FEAR OUR PEEKING INTO YOUR MENTALITY. WE KNOW WHO OTHER PEOPLE ARE IN YOUR LIFE, AS WE CAN HONE INTO THERE ENERGY PATTERNS, ONCE THEY ARE IN YOUR FREQUENCY. IF THEY ARE FROM THE EXCELTA GENETIC POOL, THEY RADIATE A VIBRATION KINDRED TO OURS. OTHERS ARE MORE SUBTLE,

206

TRANSPARENT BUT STILL READABLE. SOME THOUGHTS ARE LIKE WATER, THEY EITHER WASH AWAY OR BUILD UP LIKE A LAKE. AS THEY PILE UP OR INCREASE, THEY ARE STORED AND WE CAN READ THEM AT OUR LEISURE. SINCE YOU ARE ONE OF OURS, WE CAN EASILY PUT OUR THOUGHTS, OUR MIND, OVER YOURS. WE HAVE SYNCHRONOUS VIBRATORY LEVELS. THAT IS HOW YOU ARE NOW RECEIVING OUR THOUGHTS AND THAT IS ONE OF THE THINGS THAT MAKES YOU SPECIAL ON YOUR PLANET. DO NOT THINK THAT YOU ARE CRAZY, YOU ARE NOT. YOU ARE JUST OUT OF THE ORDERLY ROW OF MENTAL SOLDIERS ON EARTH. KEEP PRACTICING WITH US AND ACKNOWLEDGING US AND YOU WILL REACH VERY HIGH ATTAINMENTS, BEYOND YOUR CURRENT COMPREHENSION. IT IS YOUR CHOICE; WE WILL CONTINUE TO LOVE YOU EITHER WAY. WE KNOW HOW DIFFICULT IT IS TO BE OUT OF THE ORDINARY ON EARTH. IT IS NOT REWARDED, UNLESS YOU DO SOMETHING THAT IS ACKNOWLEDGED AS GREAT BY MOST. THEN THE OUT OF ORDINARY BECOMES ACCEPTABLE. DON'T LET YOUR EGO KEEP YOU LIKE EVERYONE ELSE.

YOU DO HAVE A LACK OF PATIENCE IN THIS LIFE EXPERIENCE. HARNESS PATIENCE TO RIDE THE HORSE OF THE FUTURE. IT WILL LEAD YOU TO THE STARS. YOU ASKED ABOUT YOUR POSSESSIONS THAT WE ARE HOLDING. THEY HAVE YOUR VIBRATORY "FINGERPRINTS." THAT MAKES US FEEL CLOSER TO YOU...WE DO MISS YOU SOMETIMES, ATENA. MANY OF YOUR FRIENDS ARE OF THE GENETIC STRAIN OF EXCELTA. YOU HAVE GUESSED SOME AND ALSO HAVE MANY WHO HAVE

PASSED THROUGH YOUR LIFE, AND OTHERS WHO ARE STILL TO COME. BE AWARE AS THEY ENTER INTO YOUR SPHERE THROUGH NATURAL PROGRESSION. WE WILL GIVE YOU A LITTLE GIFT TODAY, SOMETHING THAT YOU CAN USE ON YOUR OWN WORLDLY TERMS. TAKE YOUR COINS AND GO TO A SLOT MACHINE. WE WILL GIVE YOU A MIRACLE WITH YOUR LAST COIN. WHEN THIS OCCURS, TELL YOUR FRIEND LINDA ABOUT US. BRING HER BACK TO THIS HOTEL AND WE WILL GIVE YOU EACH SOME FUN. DO NOT THINK THAT WE ARE HERE TO REGULATE YOUR PROSPERITY. WE KNOW THAT YOU ARE SO RATIONAL, THAT YOU WILL LOOK FOR LOGICAL EXPLANATIONS FOR WHAT WILL HAPPEN TO YOU IN LIFE FROM NOW ON. WE ARE DOING THIS BECAUSE WE WANT TO GIVE YOU A BIT OF A PUSH IN THE DIRECTION OF EXCELTA CONSCIOUSNESS AND ALSO BECAUSE WE LOVE YOU. ALL IN LIFE IS NOT SERIOUS SO HAVE FUN AND KNOW THAT WE HAVE ALWAYS BEEN AND ARE NOW WITH YOU.

As I read back my messages from Excelta it was extremely interesting but just as was predicted, I gave myself rational explanations for what had just occurred at this table in the Sands Hotel Bar. I was having a hard time giving this experience credence. The information for the most part was good metaphysics. Since I knew nothing about mental telepathy, I couldn't validate it one way or the other. This was all new to me. I knew that I had high levels of E.S.P., but was that the same as telepathic channeling? Og from Excelta promised a gift via the slot machine. I certainly was game to try that. I had already planned to do some gambling, although Black Jack was my favorite game. My ex husband always cautioned me about the slots, saying the slots had highest odds for the house; hence you have the least chance at winning. I had also found them to be quite addictive. Once I started playing a certain machine that felt right, I got territorial and didn't want anyone else to play on it, until I got a jackpot! I saw the same pathological syndrome in other gamblers, so I normally chose to stay away from slot machines. This time, my hand was forced by Og.

I walked over to the slot area and perused the existing open machines and the various kinds of slot games. I decided to play with quarters. Dollars go too fast, especially when you are putting four in at the same time. I knew that was the best method to get a high jackpot. I bought $20 worth of quarters and found a machine that "felt right." I mentally asked Og if this was the machine. In my mind's ear, I heard, "Yes." Using mostly four quarters at a time, I got a few small jackpots but after putting the quarters back into the machine in search of a bigger one, within 20 minutes, I was down to my last four quarters. I was now beginning to doubt this telepathy bit, nothing special had occurred, out of the ordinary. I took the four quarters in my hands, closed my eyes and spoke mentally to Og. "Okay, Og of Excelta, if you are really who you say you are and I am who you say I am, I need to have a jackpot now! These are my last four quarters, so here goes...!" I deposited the quarters in the slots and three sevens came up! It *was* a winner! I won $250! I was delighted and shocked! I did not expect that to happen but now I was considering the odds of this being a coincidence. Despite my own super-rational point of view, I had to consider this to be out of the ordinary, especially since it was my last four quarters and they won! Now, I had to follow directions and tell my friends Sandy and Linda about this.

I returned to the hotel and both of my friends were in the room talking about the fascinating men they had met playing roulette. I told them about my experience in the Sands Hotel and my remarkable jackpot that was predicted by my newly found space brothers. Linda and Sandy were fascinated and wanted to know more about this new connection and how to also get connected. I didn't want to tell Sandy that Og didn't include her in a jack pot gift, so I was looking for the right opportunity to get Linda alone. I didn't have to wait for long, since Sandy said she wanted to go down to the pool and get some sun. Being late in August in Las Vegas and over 100°, fair skinned Linda declined to join Sandy. After Sandy left the room, I told Linda what Og had said about giving her a gift also. Immediately, Linda said with delight, "What are we waiting for? Let's go to the casino!" She was amazed at the fact that she was known to cosmic beings. I told her what I had channeled; if she was in my vibration, they could pick up hers and heard conversations telepathically. We walked over to the Sands Hotel and entered the slots area. "How much should I play with and which machine should I use?" she asked me. "Look, I don't know if anything is going to happen. Don't hold me to this. I just had to tell you, because of what happened to me. Otherwise, I never would have mentioned the fact." I replied, still a Doubting Thomasina.

I didn't want to push my luck, so I played the nickel and dime slots. My fingers were black with residue from the coins. I managed to pick up another $25.00 in jackpots, which is pretty good at those levels! Linda decided to take a chance and played the dollar slots and in a very short time, won a $200 jackpot. She was thrilled. I was still shaking my head at this occurrence. I tried to figure out how space guys could do this. If they weren't here in the physical, how could this happen?

Then I remembered what had happened to me on safari in Kenya and how my stuff was teleported back and forth. Exceltans obviously had the ability to manifest through the dimensional levels and manipulate our physical reality. At my current limited level of scientific and metaphysical understanding, these were concepts too difficult for me to grasp. I decided that I would just accept them as a given, like I do the laws of electricity and gravity. I might not understand exactly how they work but I have faith enough in them that I flick the switch on my electric lights and expect that they will work, and I don't jump off my balcony because I know gravity will pull me down and I will kill myself! Faith and acceptance were what I would need to enter into this communion with the cosmos. What harm would it do for me to try and see how far this could take me?

I never guessed then that this was the beginning of a life altering experience that would jettison me to a total new realm of being in this incarnation!

"When one submits himself to spiritual growth, one must be willing to travel the path through pain, sorrow, love, happiness, joy and light, and have the courage and wisdom to face himself in the image of life. So goes the warrior into seeking insight and truth becoming one with the universe because there is nothing else."
– Zarg of Excelta

"Let the light of your mind become a flame in your heart."
– St. Vincent dePaul

AND THEN THERE WERE MORE

The most difficult aspect to convey of Zarg's and the other Exceltans' transmissions is the emotion and deep feeling attached to their words. It is truly as if he and the others who speak with us are looking us in the eye, sitting across from us or embracing us! Their emotions and feelings are energetically directed. We sensed not only their seriousness of purpose but their fun and games as well. Each of the personalities we were introduced to was well defined, each of the entities had their own particular "flavor." There isn't anything wishy washy about Zarg. Everytime I spoke with him, I wished that it was possible for me to pinch his cheek, hug him and even give him the rasberries! He always has assignments for me. I am the work horse of the Excyles. From the first transmission in 1988, my assignment was to write a book about the Excyles. Zarg said, "ALL WILL COME TOGETHER WHEN YOU COMPLETE THE BOOK ABOUT THE EXCYLES. IT IS A GREAT PORTENT. GOD IS PRESENT EVERYWHERE. LISTEN TO YOUR CLOSE FRIENDS IN SPACE FOR INFORMATION. BIG POWERS ARE BEAMED DOWN IN YOUR DIRECTION. THE LIGHT WILL SHINE SOON AS YOU ARE CLOSE TO YOUR HERITAGE. IT GIVES YOU STRENGTH AND ADDS MEANING TO YOUR DIRECTION. THERE ARE TASKS FOR YOU ALWAYS, AS YOU REPORT FOR EVERYBODY HERE ON EXCELTA. MIA IS THE LOVER TO THE UNIVERSE. KEEP WORKING FOR HAPPINESS AND TRY TO REMAIN STRONG AND STUDY INDIVIDUALLY. BE MORE SERIOUS WITH YOUR GIFT OF COMMUNICATING TO MANKIND AND YOUR EXCYLE FAMILY ON EARTH. I CANNOT TELL ALL AT THIS TIME BUT WITH GREATER UNDERSTANDING, WE WILL BECOME CLOSER. OTHER EXCYLES HAVE LOST GOOD CONTACT ALONG THE WAY, YOU HAVE ALWAYS BEEN CLOSE TO YOUR UNIQUENESS, DESPITE YOUR ENVIRONMENT AND CIRCUMSTANCE ON EARTH.

THAT IS WHY YOU HAVE MEMORIES OF YOUR EARLY CHILDHOOD EXPERIENCES AND HAVE RECOGNIZED THAT YOU HAVE UNUSUAL PSYCHIC ABILITIES. YOU HAVE NOT TOTALLY SUBLIMATED THESE IN FAVOR OF WHAT EARTHLINGS CALL NORMALCY. THERE IS STILL MUCH WORK TO DO, BUT WE ON EXCELTA HAVE FAITH YOU CAN DO WHAT NEEDS TO BE DONE, DO NOT DENY YOUR SELFHOOD IN FAVOR OF THE EASY WAY WHICH BECOMES NOT MUCH MORE THAN TREADING WATER. WHAT IS THE POINT OF LIFE WITHOUT FORWARD MOVEMENT? KEEP THE MOMENTUM OF FORWARD MOVEMENT AND YOU WILL BE ENERGIZED LIKE A ROCKET AND THE ONLY THING THAT WILL STOP YOU IS THE LANDING PAD YOU ARE AIMING FOR."

How do you know when we want to talk to you?

"YOU COMMUNICATE WITH US BY MENTAL FREQUENCIES. YOU ARE HERE DEAR, WITH US IN THE ROOM IN SPIRIT. YOU MOVE TO WHAT YOU CALL THE OTHER SIDE, INTO THE LIGHT. IT IS YOUR SPECIAL ABILITY THAT ACCOMPLISHES THAT, AND THAT IS WHY THE EXCYLES NEED TO HAVE YOU EITHER ON THE BOARD OR IN THE ROOM TO HAVE ACCURATE TRANSMISSIONS. YOU ARE THE ANTENNA, THE FREQUENCY FOCUS THAT FACILITATES THE COMMUNICATION. OTHERS CAN LEARN TO ACHIEVE IT BUT IT TAKES DEVELOPMENT AND FAITH. ARE YOU AWARE WHEN I OFTEN TALK TO YOU, IT IS LIKE A WHISPER IN YOUR EAR? SOME DAYS, I WALK WITH YOU, I PROJECT MYSELF TO BE CLOSE TO YOU. WE ALSO WATCH THE EXCYLES ON OUR VIEWING SCREENS."

Can you speak to us from all the places you go to on board the ship?

"YES, WE MOVE CONSTANTLY, ALWAYS LOOKING FOR THE BEST PLACE FOR TRANSMISSION TO ALL OF OUR BEINGS AS THERE ARE EXCYLES THROUGHOUT YOUR SOLAR SYSTEM AND OTHER GALAXIES."

Where would they be in our solar system, I thought Earth was the only planet sustainable for life?

"THERE ARE THOSE WHO LIVE BELOW THE CRUSTS OF PLANETS AND NOT ALL ARE IN YOUR DIMENSIONAL ASPECT. SOME LIFE FORMS ARE NOT AS EVOLVED AS EARTHLINGS AND SOME ARE IN DIMENSIONS YOU CANNOT SEE. REMEMBER ONE SIMPLE FACT YOUR EARTH SCIENTISTS SEEM TO FORGET, JUST BECAUSE YOU DON'T ACKNOWLEDGE THE EXISTENCE OF SOMETHING, DOES NOT MEAN IT DOESN'T EXIST. THERE IS A HIGH LEVEL OF MANY KINDS OF DENIAL ON EARTH. IF IT IS IN YOUR MIND, EITHER IT ALREADY EXISTS OR IS CREATABLE, OTHERWISE YOU WOULDN'T HAVE IT IN YOUR MIND! THOUGHTS CAN EASILY BECOME THINGS, INCLUDING BAD THOUGHTS! NEVER FEAR THOUGH, GOOD ULTIMATELY TRIUMPHS, IT JUST DOESN'T ALWAYS APPEAR THAT WAY."

How is travel accomplished to other dimensions?

"IT IS EASY THROUGH SPACE WINDOWS. THERE IS CONSTANT MOVEMENT OF ENERGY WAVE RIPPLES. LIKE YOUR SURFERS, WE WAIT FOR THE RIGHT CURRENT AND RIDE OUR SHIPS IN ON THE RIPPLE. THERE IS AN EBB AND FLOW IN THE ACCESS WINDOWS. THE BERMUDA TRIANGLE IS ONE IN THE FLORIDA AREA. THEN OF COURSE, WE CAN COMMUNICATE INTERDIMENSIONALLY THROUGH BRAIN WAVES, WITHOUT CRAFT. WITH CRAFT, WE OFTEN SLIDE THROUGH GALAXIES TO MEET WITH

215

OTHER MEMBERS OF THE ASTRAL ASSOCIATION TO COMPARE NOTES AND CHECK OTHER STAGES OR LEVELS OF CONCIOUSNESS FOR DEVELOPMENT."

How did the Excyles enter Earth bodies?

"THE SOUL CHOOSES THE ENTITY FOR THE EXPERIENCES NEEDED TO PROGRESS OR BALANCE KARMA. AS I HAVE MENTIONED BEFORE, EARTH HAS A VAST OPPORTUNITY FOR EXPERIENCES NOT AVAILABLE ON OTHER PLANETS AND DIMENSIONS. EXCYLES HAVE BEEN ON EARTH SINCE ANCIENT TIMES, THAT IS HOW THE CIVILIZATIONS BECAME SO ADVANCED. YOU ALL HAVE BEEN AROUND EARTH DURING THESE PERIODS AND HAVE ESPECIALLY EFFECTED ATLANTIS AND EGYPT. THERE ARE E.T.s VISITING EARTH TO LEARN MORE ABOUT YOUR PHYSICAL BODIES WHICH ALSO HAVE UNIQUE COMPONENTS, ESPECIALLY YOUR HANDS AND FEET AND GENITALS, MANY FIND THEM EXTREMELY INTERESTING. DO NOT FEAR THAT THESE ENTITIES WHO USE EARTH PHYSICAL BODIES FOR THEIR SCIENTIFIC INVESTIGATIONS AND PURPOSES HAVE DOMAIN OVER YOU PERMANENTLY. THE POWER FOR GOOD RULES ALL THINGS AT ALL TIMES. LOVE TRIUMPHS ALWAYS OVER EVIL. LOOK AHEAD FOR ALL GOOD THINGS LIE AHEAD. REACH FOR A HIGHER PLANE. LET LIFE FLOW AND YOU WILL SUCCEED IN YOUR ENDEAVORS. ALL WILL HAVE A HARMONIOUS OUTCOME AS EVERYTHING IS BEING DONE THAT IS POSSIBLE. MANY LEVELS OF CONSCIOUSNESS ARE HELPING AND WORKING FOR THE GOOD OF MANKIND. IT HAS ALWAYS BEEN AND ALWAYS WILL BE, THAT IS THE NATURE OF THINGS. KEEP IN MIND THAT IN RETROSPECT, EVERYTHING IS IN ABSOLUTE AND PERFECT ORDER IN THE DIVINE PLAN. LEARN TO

216

TRUST YOUR OWN MIND VOICE AND DO NOT BE DISTRACTED BY APPEARANCES OF THINGS. THERE IS ILLUSION IN DISILLUSION. YOU MUST HAVE FAITH. THE MASTER MUST PICK A DIRECTION FROM HIS HEART AND FOLLOW IT TO THE END."

As the years progressed, just as Zarg had prophesied, our Excyle fold increased. Some came into the fold in great wonder and astonishment and others in doubt and smirking cynicism. For a few years, we had a yearly Excyle reunion but I got tired of having to push the buttons of togetherness. I felt either they were attuned to the truth of their being or they were not. There were a few whose belief and enthusiasm never wavered. Of course, Troy, King Dugud, whose personal life was often beset with travail, always kept the faith with a resiliance and resoluteness of purpose, of which I was in awe. He reminded me of those punchbag toys that you pushed down and always popped back up again! That was Troy, "No wonder he is the King!" I would often say to myself.

One thing I noticed about the Excyles I knew, even those who didn't take their heritage to their bosoms, almost all moved their lives forward in a very positive and growth producing way. Several returned to college and others improved their careers and businesses, which produced much happier, focused and centered lives.

I believe the greatest benefit has come to me. I have never felt happier or more alive, and this feeling increases every day. As Pollyannaish as it sounds, it is true! Some people are troubled because they don't know what their "mission" is in their life. They continually are on the prowl for something or someone to make them feel fulfilled or worthy.

Even though I was a happy and accomplished human being before 1988, since Zarg entered my life with his revelations, I feel truly born again! I have been baptized by the spirit of inner knowing, that part of me that never lies to myself no matter what. It has become a sacrement to me signifying my spiritual rebirth, not just in my connection with Excelta but with my reunion with my Cosmic Parent, the Eternal Energy that is the One. In this initiation rite, I have released most of my fear of the Unknown, the dark morass that threatens us all, of what becomes of us once we drop our bodies. We are not bodies with souls, we are souls that inhabit bodies! I know for sure that life sometimes seems like a baptism of fire but is really just a basic initiation rite. Over and over the pervading theme that Zarg has presented is that life just asks us to do two things, love and learn. How much you love and how much you learn is the water mark of lifetime evolvement. That is the simple and most difficult mission for each soul every life experience.

It is a universal corollary. It is the ordeal that tests our endurance best. No sentient being that God creates can outwit the process, we all are involved in it, even at the 22nd Excelta level. There is no end to spiritual evolvement, even at the Godhead! The mind that created us and that recreates us must then recreate Itself, always in the process.

Life is our mission, anything else is peripheral. As Zarg was instructing me so well in my lessons, I found the strength and courage to use the truths he shared. As my life evolved, I had plenty of opportunity to test my metaphysical mettle. In my search for more knowledge about cosmology and ufology, I had been drawn into the the intrigue of the United States Intelligence Agencies fun and games. If ever there was a test, that is it. I believe I have weathered it well.

Jordan, FBI Agent extraordinaire, said the U.S. Government has known about my contacts since my childhood. Even though I don't need Jordan's validation of my experience with Zarg, I therefore can assume that the information I have received is correct. Any kind of spiritual advice that is centered around loving God and mankind can't be wrong and can't do any harm!

For those of you who would admonish me or anyone else for using a Ouija board or giving credance to any information emanating through it, I would ask that you open your minds a bit. The use of Ouija-like devices is not new. In ancient civilizations, mystics used gadgets to call upon wisdom from the unseen realms. Jane Roberts contacted Seth, an "energy personality essence, no longer in physcial form." Seth produced through Jane Robert's Ouija board, incredibly detailed spiritually based information. Similarly profound information was given to Jessica Lansing via the entity Michael. Books and poetry have also been dictated through Ouija boards.

In the past, I have had a great distaste for the use of what is now commonly called "channeling." I looked askance at it, as much as I had the use of the Ouija board. I felt that the information channeled was colored by the receiver. If the channel was not clear and centered spiritually, most likely the information was "tainted" by their belief systems and prejudices.

There is no doubt in my mind, that I was channeling Zarg and others from his level. Even though the planchette often moved quicker than my mind could tabulate its meaning, I was the conduit for the energy. I was the receiver. The force that moved it was through me via mind link. If it was at a subconscious level, it came from information I had culled through the milleniums on my way to Excelta and at the level of Excelta.

As Zarg stated, the use of the board was my choice, an easy way out in the beginning. At the level of third dimensional reality, the Ouija board acts as a physical focus for what is travelling through the Mind of God, a fluid medium of mental energy that acts as a conductor of information. I know now that I no longer truly need the Ouija board. I can and have received information directly on a mind to mind link from Excelta, the difference is that I trust the information more. Therefore, I am not really any different than those channels I eschewed, but I know how clear the channel that I am is! I know the mind set and the emotional baggage I am carrying. So, I can be better assured of the validity of the information for myself, other Excyles and those with whom I share the information.

I hope that I can live up to Zarg's expectations of me. His words ring through my being as inspired challenges and goals.

"MIA, SHOW THE LIGHT BY BECOMING BETTER AND LEAD THE WAY FOR OTHERS TO SEE HOW PERFECT LIFE CAN BE BY BEING GOOD. WITH GREAT LOVE FOR OTHERS, PEOPLE WILL WONDER AND MARVEL AT THE BEAUTY OF YOUR SOUL. WITH JUST A LITTLE EFFORT, YOU WILL SEE IT HAPPENING EVERY DAY. YOU MUST BE AWARE AND SUPPORT AND GIVE COMFORT TO ALL THE EXCYLES AND OTHERS. AS YOU GO ABOUT YOUR BUSINESS OTHERS WILL LEARN FROM YOU, AS YOU RADIATE BY YOUR EXAMPLE. SOME WILL FIGHT THE WAY BECAUSE IT IS NEW FROM WHAT THEY KNOW. YOU MUST HAVE FAITH. THE MASTER MUST QUESTION EVERYTHING TO CHOOSE THE CORRECT DIRECTION. PICK A DIRECTION FROM YOUR HEART AND FOLLOW IT TO THE END. KEEP THE CHANNELS OPEN AND REMEMBER WHO YOU ARE AND WHERE YOU ARE FROM. FOLLOW THE LIGHT TO THE END OF THE TUNNEL WHERE THE TRUTH WILL REVEAL ITSELF. WE CAN POINT IN THE DIRECTION BUT YOU MUST DO THE WALK. LOOK AHEAD, DON'T LOOK BACK, BUILDING FOR THE FUTURE AS YOU GET CLOSER, WHILE ALWAYS KNOWING HOW POWERFUL YOU ARE."

REUNION IN KEY WEST

There is no place like Key West, at the southernmost point of the Florida Keys. Throughout its history, it has maintained its reputation for Pina Coladas and rugged individualism. It is an amalgam of people, cultures and wonderful smells wafting from great restaurants. I love it! What could be a better...*and* worse place to do a juice fast? I decided to anyway. Russell House, situated in a remodeled motel, had a program to loose weight drinking juice water, tea and other non-delicious things, but it also got results. I needed to loose 10 pounds and relax, so down to Key West I went.

Before you do anything in the morning, a fast one-hour walk with other juice waterers was mandated. The first morning I was there, it was led by Daisy, a soft spoken, red-haired young woman. Her hair was truly her crowning glory, as it fell in long, thick ringlets, like one of Botticelli's women. There were a dozen of us tromping through the side streets of Key West, passing shop keepers getting ready for a new day. They were used to the rag tag band of tubby tummies trying to look like they were enjoying the agony. I was near the head of the pack, huffing and puffing away. Daisy said hi to me and introduced herself. She asked my name and I responded. Then totally out of the blue, she asked me a question. "Mia, do you know anything about dreams? I have been having these strange dreams and I would like to know what they mean." I told her that I knew a little bit about dreams and asked her what they were about. "I have been dreaming that I am on a space ship and am looking out of the window. I think that I am at the helm, since it is a big window, like a picture window. We are moving along at quite a fast pace and are heading for a planet that feels like it is home. When I look at the planet ahead, I feel like it has a heart around it, and I assume that I am connected to it. I am always about to arrive but I feel frustrated because I don't seem to be able to get there. There is a man with me and we both are staring out longingly at space."

Daisy continued, "He is pointing towards the planet with the heart around it. I am standing at the window with daisies in my hand." She stopped speaking and looked towards me out of the corner of her eye. "When I first saw you this morning, I had the feeling that you could tell me about my dream."

I was amazed at what I was hearing so early in the morning and before I even had my juice water! Now what do I do? Do I tell someone I don't know, the spatial facts of life. I knew exactly what her dream meant, but did I dare tell her? I didn't even know her! But then again, she did pick me out of all of the rag tag plodding alongside her! She must have had some gut feeling for herself. Then I recalled, in my last communication with Zarg, he told me that I would have a surprise when I went to Key West. I didn't think that the surprise would come in the form of a hippie looking Botticelli beauty. I was thinking more of a conch Key Wester macho man! I asked Daisy if she knew anything about UFOs or extraterrestrials. She said she didn't but that she always was fascinated with the notion of them. I told her that I would rather discuss the dream when we weren't moving faster than a speeding bullet. She nodded in agreement and we set a time when we both would be available to discuss her dream and some other things I wanted to share with her.

I knew either intuitively or via telepathy from Zarg that Daisy was an Excyle. Obviously the planet with a heart around it was Excelta, or it represented our home star system. I believe that the man standing beside her represented another Excyle. Since she was holding daisies that implied that it was her lifetime on Earth now. Perhaps this lifetime she either had the opportunity to return to Excelta or to know who she really is, a cosmic visitor. The fact that she sought me out amongst a bunch of people she didn't know was very significant. Somewhere in her soul memory she felt a connection with a kindred spirit. Also, Zarg did hint that I would have something occur that was important.

I met with Daisy and shared my evaluation of her dream and also some of my connection with Excelta and other Excyles. I didn't want to say for sure that she was, but I said that I would check with Zarg. In my heart and mind, I already knew the answer. I could also mentally see Zarg's smiling face and I knew that all of those who are involved with us here on Earth were happy that one more Excyle was connected with the others. I began to clearly see that being the connecting link was the role I was chosen to play. Whoever was penning the script and directing the scenes was doing a good job!

Zarg did verify Daisy's connection to the rest of us. He would remind me that I was to keep in touch with her and would tell me things about her that he thought I should know. She moved from Key West to a north Florida town. She fell in love with an abusive guru type and got pregnant with his child. Zarg told me it would be a very special male child. Daisy was determined to get this guy to change his ways, but he was just as determined not to. She would phone me for some counsel but did not like what I had to say. I hoped her story would have a happy ending.

Zarg did reveal something else interesting about Daisy. While Daisy was still living in Key West, and after our meeting, she told people about our conversation and about the Excyle connection. Unfortunately, she told the wrong people. Key West is loaded with lots of government agents due to its proximity to Cuba and its temptations for drug traffickers. Apparently, that was one of the reasons that some G-men found me so interesting. Most of the intelligence agencies have special bureaus of UFO affairs. I became another name in their computers, and so did Daisy, but these were not the only government types to focus their hi-tech hijinks on me in psychological games without rules or boundries.

The building itself has an anecdotal history of attracting other-worldly beings due to the dimensional doorway that opens up, as well as its physical location in the Mojave that has attracted interdimensional intelligences for some time.

INTEGRATRON INTENSIVE

My time spent juice fasting in Key West not only provided my introduction to Daisy, it also provided an opportunity to go to the Integratron. I was sitting around the pool at Russell House, trying to keep my mind off of my growling tummy, when I picked up a new age focused newspaper. An advertisement caught my eye and excited my interest. It was for a group intensive retreat that was going to the Integratron in the Mojave Desert of California. The main statement that titillated my curiosity was that the Integratron was a building built by George Van Tassel with instructions from interdimensional beings. The purpose of the building was for integration of our spiritual consciousness with our physical and mental bodies. The name Integratron was given to Van Tassel by the beings. The retreat sounded fascinating, so I phoned for more information when I got home.

The retreat was held in August. There are no sleeping facilities at the Integratron, so we were told to bring a sleeping bag and prepare to sleep in the desert near the building. It was a good thing that I was a Floridian because the Mojave in mid summer is not for wimps! There were no cooking facilities so we brought whatever was edible that did not need cooking. Thank God there was a refrigerator for cold stuff...we needed it! We slept in tents. I shared mine with a man and woman.

There were 15 of us communing and "OM-ing" and hoping to avoid the rattlesnakes. Zarg had assured me that there would be none and, happily, he was right.

George Van Tassel was a former test pilot for Howard Hughes and Douglas Aircraft. He said that was walking in the Mojave and came across interdimensional beings who gave him a mission to build the Integratron. The Integratron is not just a non-metallic building, it is a machine. It is a high voltage electrostatic generator that supplies a broad range of frequencies to rejuvenate the physical body and recharge the cell structure. The building itself is unusual in that it is a completely non-metallic, domed structure approximately 50 feet in diameter and 38 feet high. It uses vortexian mechanics and the mechanics of physics in producing a sound chamber. Inside the domed structure, the sound phenomenon produces a way to redevelop the brain into a whole-brain relationship, not just left or right brain dominant. In so doing, it facilitates movements interdimensionally, which is the next step for planet Earth. Whole brain integration is said to move individual consciousness and therefore allows access to higher interdimensional realms. There is the added benefit of cell rejuvenation, which could reverse the aging process, or even stop it.

Inside the dome, besides using the "OM" sound together, we also used bells and crystal bowls to produce additional resonance. It was truly entrancing. The place makes any sound system seem state of the art. We also played rock music and danced. Getting loose and feeling like a kid again is part of the process, as well as a lot of fun. We didn't have to play the music or sing loudly, the building did it for us. Even the most subtle whisper could be heard clearly 25 feet away. We were told that we must challenge the tympanic membrane to regenerate the nerves in the left and right brain. Each participant was circled by others chanting, "OM-ing," ringing bells or using bowls. The central person became the focal point for the vortex and healing was said to commence.

The building itself has an anecdotal history of attracting other-worldly beings due to the dimensional doorway that opens up, as well as its physical location in the Mojave that has attracted interdimensional intelligences for some time. George Van Tassel and others in the 1950's and 60's had numerous UFO sightings in the Mojave near the Giant Rock and Crystal Mountain areas. Geologically, the area is a solid bed of quartz crystal with underground rivers that run from there to Mt. Shasta in north-central California. Mt. Shasta also has a history of paranormal activity with space craft. It would seem that the connection of these two areas creates a vortex, possibly ley line activity similar to England's.

Van Tassel completed his 18 year building project of the Integratron and then died mysteriously. The subsequent history of the building is fascinating. The U.S. government came in and removed the mechanical and technological equipment that was made specifically for the Integratron. The equipment's whereabouts are unknown. The building was held by the government for several years, then later sold and made into a restaurant and disco. Eventually, the Integratron was rescued by a metaphysical teacher who has allowed it to be used only for its intended purpose of evolving consciousness.

Personally, my greatest sense of connection was not the building, but the environs. We trekked to the Giant Rock, Crystal Mountain area to do some meditating. This place was the site of UFO conferences in the 50's. Many of the pioneers of ufology and the early abductees and contactees had their platforms to tell their tales there. Now it is just a footnote in history. The area is very special. I felt its power immediately. Crystal stones are found all over the area. After a few minutes of my arrival, I heard in my mind, "This is where you were hatched." It seemed like there was kind of a giggle with those words. I had an overpowering feeling of communion here. I know why the word "hatched" was used. It was meant as a double entendre. I left through the "hatch" of a spaceship from my home star and landed on to this planet thousands of years ago. I was also like a chick being "hatched", since I was a infant arriving into a new world, Earth! Edwards Air Force Base, where our space shuttle lands also is near this area! Another coincidence!

Coincidence or not, perhaps there is something in the consciousness of humans that tells us the appropriate places to leave and return to Earth, and maybe the military knows the appropriate points of easy entry and egress off planet.

This area of the Mojave has undergone massive geological changes through the eons of time. The Lemurian civilization of the super-ancient period might have situated itself nearby, if not right there, due to the crystalline structure that was used extensively in that and the Atlantean eras. I feel sure that I have some connection there, since my skin tingles just thinking about that area. It is interesting to me that the affinity is so great, since I don't want anything to do with the rest of California, except Mt. Shasta. Since Mt. Shasta and the Mojave are connected energetically, I can understand my attraction. Zarg verified my sensings at Crystal Mountain.

I am glad that I had the opportunity to visit and experience the Integratron. It seems that I am always led to the places I need to experience and in this case, re-experience.

LAKE TAHOE ENCOUNTER

In December, 1988, Troy, Dan, Dan's friend Chip and I went for a ski holiday to Lake Tahoe. It was a glorious week. There was a record snowfall that month and Tahoe ski areas were extremely busy with snow-bums and bunnies from the surrounding metropolitan areas. We had no complaints about the spectacularly pristine and deeply powdered ski turf. Troy was in skiers' heaven. His times spent on both the Nevada and California sides of the ski areas were quite productive. Due to the perfect ski conditions, he made quite a lot of progress in his downhill technique.

The last day that we had to spend in Tahoe before we flew home, Troy suggested that he and I have a special going away lunch in the restaurant on top of the Heavenly Valley Ski Lift. It had a panoramic view and excellent food and would make a fitting ending to a special week. We were to meet there at noon.

I made my way up via cable car and admired the most spectacular view. When I got up to the top, I noticed how crowded the dining and snack areas were. Humanity clomping around in snow boots created an awful din and a hazard on the slippery stairs and walkways. I walked through the snack areas filled wall to wall with hungry, pink-cheeked skiers and elbowed my way to the restaurant. There was a line of hopeful diners waiting to get in, and after their conversations with the maitre d', I noticed that they either moved to the bar to wait or shrugged their shoulders and walked away. But I thought that I would take my chance anyway and pushed to the front of the line. When my turn came, I asked how long it would take to be seated. The maitre d' gave me a sour expression and said, "It is a two-hour wait and the kitchen is quite slow today!" He certainly made no attempt at being encouraging. I knew that Troy wanted to make use of every minute he had left schussing on the trails and the food situation did not look promising, so I, too, shrugged my shoulders and walked away.

When Troy arrived, he asked me if I had made reservations. I gave him an overview of the situation and suggested a quick bite in the snack bar. Troy, being the original "never say die guy," said that he was going to try. I thought that he was just wasting his breath as I (and dozens of others I'd seen) hadn't had any luck getting in. Troy clomped over in his ski boots towards the maitre d'. Much to my astonishment, when he asked about lunching, the maitre d' said that it was busy but he would try to find something. I looked closely and NO cash passed hands. Troy suggested that perhaps someone wouldn't mind sharing a table. At that, the maitre d' said, "Oh, yes, I believe I can find someone who will." Almost instantly, he came back and eagerly ushered us to a table for four that was solely occupied by one man. I thought, "I don't know what kind of magic Troy performed but it surely did the trick."

I was still shocked as I sat down, but quite happy. Both Troy and I were bubbling over with enthusiasm at the mountain view and the great stroke of luck to get a table with no trouble or waiting at all. We introduced ourselves to our lone lunch partner. I wish that I could remember his name but I cannot. The man was quite unusual looking. He had straight, shoulder length, sandy brown hair and a deeply tanned, angular face and the bluest of blue eyes. With the deep tan, those eyes practically jumped out at you! He had an ageless quality about him. He looked fortyish, but he could have been fifty or even more. It was hard to tell, especially because of the shoulder length hair. That length had gone out of style so it was unusual to see it, especially on what appeared to be an "older" person.

We introduced ourselves and told him that we were from Florida. I asked him where his home was and he said, "Texas." I probed further and asked him what city in Texas. I will never forget the look on his face when he responded, "All over." He said this with a kind of a shy smile that made his eyes appear to twinkle. "Well," I thought, "this guy likes to be secretive. Perhaps, he's an oilionaire who has ranches and homes all over the state and he's trying to pass himself off as a regular person!" My mind boggled with the possibilities. He also was sitting by himself at a table for four in an overcrowded dining room in the height of the ski season. That in itself was a remarkable feat!

The man did tell us that he came to Tahoe to ski the season but did not reveal much else. He was very quiet for someone who was happy to share his table with strangers. We ordered our meal and it was served to us quickly, despite the fact that the kitchen was supposedly slow, as was told to me earlier. Everything was going perfectly. Troy was waxing poetic about his morning on the slopes and spoke of his feeling of oneness and attunement to nature. This also was greeted with twinkling eyes and smile from the man. Not a word was uttered from his lips until Troy and I were discussing an argument Troy had with Chip the night before. At that, the man made a short but metaphysical statement to Troy about the argument. I was quite surprised. I asked him if he was into metaphysics because his words sounded that way. He responded, "I wouldn't say metaphysics exactly...let's just say that I'm into the Great Spirit."

That statement about the Great Spirit catalyzed a perception in my mind. All of a sudden that deeply tanned, blue eyed face changed and he started to look like an American Indian. "Maybe, he's a half-breed, I thought. That does explain the Texas roaming-home bit." A short time later, the man excused himself and got up to leave. When he did stand up, we saw an extremely tall person, at least six feet six inches tall. He was wearing a very form-fitted, one-piece ski suit that showed a lithe and muscular body underneath. He sure looked good for a man who was at least middle aged. As a matter of fact, he looked great for a man of any age!

Troy turned to me after we watched the man's departure, "He looks like an...," "...American Indian," I finished. We both got that image at the same time. "Did you see how tall he was?" Troy said in wonder. I certainly agreed with him on all counts, and shared his amazement. We didn't think or talk much more about the man after that. We went on to enjoy the rest of our meal and our final day in Lake Tahoe.

Our first time speaking with Zarg after we returned from our ski holiday he made an astounding statement. In discussing our trip, he told Troy and me that the man who allowed us to share his table was one of Troy's "Guides." He had come specifically to check up on Troy's progress and was quite satisfied with what he had heard from Troy's lips. He was especially happy with Troy's communion with nature. The guide's name was Leard and was sent by the Higher Authority. Troy was allowing him to see his progress as a good human being. Leard's energy vortex was allowing Troy to mirror his progress subconsciously. Troy found his truth at the top of the mountain with God and became love incarnate there. Leard felt that Troy had learned his lessons well and was letting go of fear. The guide was most pleased at Troy's analogy about the mountain's being like life and skiing on it was part of the metaphor of experiencing life's travails.

Hearing this, I was somewhat incredulous. I had heard many metaphysical people talk about their so called Indian Guides in the astral or etheric realms. I never gave them much thought. It seemed like a lot of hooey to me. Now here is my beloved Zarg telling us that there IS such a thing and that we saw one in the flesh! I never believed such a thing was possible. It seems a lot easier to believe in E.T.s than to believe in true spiritual beings manifesting in the physical. "What did that guy think about me?" I asked. "He wasn't there for you, he was there for Troy, and Troy passed the test. You were supposed to be there as a part of Troy's life," he said. Just my luck, I have had an opportunity to meet a real live, honest to goodness spiritual being and he couldn't care less about me!

In retrospect, I just wish that I had paid more attention to the man. I was trying so hard to be polite because of his generosity in sharing his table and our incredible luck, I didn't ask more probing questions. Maybe that's the way it was supposed to unfold. Now I understood why Troy was able to get us into the restaurant and I couldn't, and why there were seats for us at this special table. My life *is* stranger than fiction!

EXCYLE REUNION

In September, 1991, we had a party celebrating the third anniversary of contact with Zarg and Excelta. A dozen of the identified Excyles gathered for reunion with other kindred energies and to speak with Zarg. He had something special planned for us. Soon after the communication began he introduced us to someone new. She is Maleha, a female Native American spirit guide whose incredible wisdom held us spellbound. Her words underscored the prevailing theme of all of our contacts that we each had a destiny in a universe that extends past the modern technology of current Earth. If we have faith in our lineage and remember who we are, many good and wonderful things will happen as Maleha's spirit will help guide and protect our way. Often since, I have felt Maleha's comforting presence and the strong, loving energy of her connection to the Excyles.

"CALL ME SUNSHINE, DAUGHTER OF RAINSTORM, LEADER AND PROVIDER OF ALL NATIONS BEFORE YOU. I GIVE YOU A GIFT, MUCH LOVE, THE UNIVERSE, THE EARTH AND GOD ARE YOU.

YOU ARE CREATION ABOUT TO GIVE BIRTH TO UNIVERSAL CONSCIOUSNESS. YOU HAVE BEEN CHOSEN FROM THE MANY. I WILL BE ALWAYS WITH YOU.

I AM THE FEMININE ENERGY OF YOUR SOURCE. THE ENERGY OF YOUR MATERNAL INSTINCTS, THE BIRTH OF A NEW NATION WITHIN YOUR OWN BEING. FAITH, TRUST, WISDOM, LOVE AND LIGHT WILL BE YOUR ARMOR.

MAKE WELL YOUR TIME. GREAT HAPPENINGS PRESENT THEMSELVES SO BE READY TO RECOGNIZE OR THEY WILL PASS BEFORE YOU. RICHES IN MATERIAL, HEALTH AND LOVE.

TIME IS NEBULOUS. WHAT SEEMS AS TIME TO YOU IS NOT WHAT TIME IS. YOU ARE TIME NOW. TRANSFORMATION ACTS ON YOUR ABILITY TO RECOGNIZE YOUR POWER TO DISPEL YOUR FATE. PRESENT TIME IS INDEFINITE. YOU ARE NOT THE ONLY PEOPLE IN A SAME MOMENT. TIME IS MULTI-DIMENSIONAL. YOU CHOOSE ON WHAT LEVEL YOU ARE. YOU CAN CHOOSE AND CHANGE LEVELS AS WELL.

THE MIND DICTATES, THE SOUL CAN BE ON DIFFERENT LEVELS. THE MIND CANNOT. ENERGY KNOWS NO BOUNDARIES. YOU NOW DIRECT ENERGY TO A GIVEN POINT YOU MA. MEET ME AT THIS POINT WHERE I SHALL REVEAL THE TRUTH.

IT IS YOUR CHOICE. I CANNOT MEET YOU HALFWAY. IT IS YOUR JOURNEY, YOUR FATE IS IN YOUR HANDS. TO LEARN, YOU MUST SEEK THE TRUTH WITHIN AS WELL AS ABOUT AND ABOVE. THE SECRETS OF THE UNIVERSE I WILL GIVE YOU.

THE SANDS OF TIME ARE LIKE THAT IN AN HOURGLASS. IT MAKES NO DIFFERENCE THAT THE SANDS ARE RUNNING THROUGH. YOU HAVE THE POWER TO TURN THE HOURGLASS OVER AND CONTINUE INDEFINITELY. THERE IS NOT END OF TIME, YOU CANNOT KILL IT, ONLY WASTE IT.

YOU MUST LEARN TO BE STRONG. LOOK IN THE CLOSET FOR A MEMORY TO EVOKE A STRENGTHENING.

YOU MUST FREE YOURSELF NOW. NO MORE FEAR, REST WORRY AGAINST YOURSELF OR ELSE YOU BECOME YOUR WORST ENEMY WITH TOO MANY QUESTIONS OF CONFLICT. IT DOES NOT MATTER WHY, JUST BE. THE DOOR IS OPEN, YOU CANNOT PASS

IF YOU STAND IN THE DOORWAY. HAVE TRUST AND FAITH IN THE UNIVERSE. YOU MUST RELEASE IN ORDER TO RECEIVE. DECIDE FOR YOURSELF AND BE WHAT YOU BELIEVE. FAITH, ENERGY AND POWER PREVAIL. PROBLEMS EXIST, IF YOU ALLOW THEM. YOU ARE THE STAR IN YOUR OWN UNIVERSE.

TWINKLE."

Death is a transitionary state until we move to our next reality, whatever that may be.

IS HE HOME NOW?

In July, 1995, I was attending the Sunday service at my church .It was the first time in 14 years that I have attended that church that the minister's topic was on reincarnation. As I expected the viewpoint was typical of our metaphysical discipline that we must live in the now of life, since now is our job, the past has lead us to this point and dealing with our reality is why we have chosen to be here. Death is transitionary state until we move to our next reality, whatever that may be. At the end of the service, as was normal, the minister thanked the donors of two bouquets of flowers on the dais. Normally, there is only one bouquet per Sunday. When she announced the donor of the second bouquet and its purpose I went into shock. She said, "The second bouquet is a special one donated by Jerry Pearson in memory of Roger Barton." I couldn't believe what I had heard, she couldn't have said "in memory of," I had dinner with Roger few months before. Roger had recently come out of the hospital because of a neurological condition but he said he was feeling much better. He looked pale but appeared to be doing well. Roger was a very private person who didn't like to be probed too much. I sensed this and kept my inquisitive distance, I just wished him health for the future.

Roger was an Excyle. He had conscious memories of an experience aboard a spacecraft during his childhood in Indiana. Roger accompanied me on one of my trips to Sedona and had surprised me with a photo album of our trip, including photos of us at the vortexes where beams appeared. Even though we didn't see each other regularly, we still had close and loving feelings when we did.

As soon as the service was over, I ran to the back of the sanctuary to try to catch Jerry as he was leaving. I could already feel the tears welling up in my eyes, even though I was praying that I had heard the minister wrong! As soon as I saw him, I asked him when Roger had died and why no one had told me. Jerry took me in his arms, as the rest of the congregation passed through the doorway.

237

"Roger made us promise that we wouldn't tell anyone that he was dying. He passed away at home. He didn't want a memorial service or any big deal. His mother has Alzheimer's and doesn't even realize that he is gone. I have his ashes and we are going to throw them in the ocean next week. We are going to have a little celebration of his life then. I will let you know when."

He gave me a kiss and another hug and then we said our goodbyes.

I was terribly sad and I felt angry. I wished that Zarg had let me know that this was going to transpire. In retrospect, I remembered several times when Zarg suggested that I phone Roger. It was not unusual for Zarg to do that, he often mentioned other Excyles who he said needed my attention. I kind of took that half-seriously, thinking this another one of the tasks for Excelta. Now, I wished I had paid better attention to Zarg's words.

Afer Jerry left, I went to lunch with a friend I hadn't seen for awhile. I was feeling a little better but still very sad. When I came home, I went into my bedroom to change into more comfortable clothes. As I sat down on my bed to take off my shoes, I noticed that a video cassette had fallen from the shelf that held my VCR tapes. I picked the cassette up and looked to see what cassette it was. There was no label on the box or the cassette. I noticed that the tape had not been rewound but was stopped in mid cycle. I had a very strong urge to play the cassette tape.

I didn't change my clothes but went straight to the VCR. it was a good thing that I was sitting down because if I hadn't I would probably have fainted. What had been recorded on the tape was a two hour PBS program on death and dying I had recorded a few years ago and had forgotten about. The focus was on how American and other cultures deal with death itself and the prospect of death. It couldn't have been more appropos for the moment I was in. By what hand and what miracle of synchronicity did this happen? Was this Zarg's or Roger's way of helping me through this realization and pain? Perhaps it was both? Hopefully Roger found his way back home to Excelta and is being embraced by our spiritual brothers and sisters. It then occurred to me that the minister's topic of reincarnation was also totally appropriate for this day. If I was grieving, it was only for my personal loss of a very dear friend and my sorrow for not having been able to say goodbye the way I would have wanted.

In James Redfield's Celestine Prophecy, he emphasizes the importance of synchronicities in our lives as valuable perspectives for the future. We become aware of the coincidences that occur and are happening more frequently and appear to be beyond chance and probability. Behind these chance occurances there is a strong feeling that something unexplainable, some kind of force is in operation to make them happen. It is when these occur and especially when they happen often, we feel connected to something Divine, something mysterious, something conscious!

Synchronicities and unexplainable coincidences are a recurring theme in my life. In Roger's death and my loss of him in the physical, the synchronistic appearance of the videotape has again reconnected us beyond our mortal friendship. From across the pale of an unseen reality, a force that cares created closure for me. In Roger's death and the appearance of the videotape, I feel enriched by the mystery of life and the forces that propel the eternal soul as it moves into the cosmic sea.

Several days later, Troy and I decided to talk to Zarg.

ZARG: I HAVE SOMEONE WHO WANTS TO TALK TO
          YOU BOTH.

TROY: I wonder who that could be?

MIA: Maybe it's General Wapet?

239

ROGER: *HI, I AM HERE NOW. THIS IS ROGER.*

Troy and I both gasped in surprise! I had tears of happiness and amazement. The planchette moved slowly and precisely, we both could feel Roger's energy.

MIA: Roger, you must have been reading the chapter I just wrote about you for my book.

ROGER: *YES. I ESPECIALLY LIKE YOUR DESCRIPTION OF EXCELTA AND WHAT IT IS TO BE HERE. IT IS APPROPRIATE. THE VIDEO TAPE WAS MY TOUCH. I WANTED YOU TO KNOW THAT EVERYTHING IS GOING WELL FOR ME. I AM HAPPY TO BE HERE. THE EXCYLE REUNION IS NOW MY JOB. I WILL BE HELPING OUT WITH THE PROJECT IN GENERAL, WE WILL BE GATHERING SOON.*

Troy and I exchanged fearful glances.

TROY: Does that mean that we are going to die sometime soon?

MIA: If so, I don't want to know when or where!

ROGER: *NO, YOU HAVE A LONG TIME YET. DON'T WORRY.*

MIA: Do you miss your dog? I know how much you love him.

ROGER: *NO, I AM AT A DIFFERENT LEVEL WHERE THOSE THINGS ARE NOT EVEN IN MY MIND. I HAD BEEN PREPARING MYSELF TO COME AND THE EXCELTA FAMILY WAS WAITING FOR ME AND EMBRACED ME WITH GREAT LOVE. THEY ALSO REMOVED THE IMPLANT.*

MIA: Which implant?

ROGER: *FROM WHEN I WAS ABDUCTED IN INDIANA.*

MIA: Did you find out who abducted you then?

ROGER: *IT WAS THE GRAYS. BUT THAT IS ALL OVER NOW. I AM ADJUSTING WELL AND AM TRANQUIL. I FEEL AS IF I NEVER LEFT OUR EXCELTA HOME. I LOVE YOU BOTH AND WILL BE WATCHING OVER YOU NOW WITH ZARG AND THE OTHERS.*

CHAPTER THREE

JOURNEYS INTO
PARANORMAL
PARADIGMS

When I protested that I was no one unusual, just an average cosmic being, she shook her head. "Oh no, you're not. It seems to me that you are special. Keep up the good work."

ENERGY ALERT

Coming from the midwest, my mother had never experienced a New England autumn. Since her arrival in South Florida, she had made many new friends including Bostonians and other New Englanders. Their tales of the glorious colors of the vegetation during the fall really piqued my mother's curiosity. She was determined not to miss out on God's palette and told me so. We discussed taking a trip northward at the opportune time for the optimal colors. After consulting travel agents, we made reservations for a New England foliage tour for mid-October.

We were to meet our tour in Boston and took a plane from Ft. Lauderdale. The plane was a wide bodied DC10-11. The flight was filled almost to capacity. We were lucky, since we had the two seats near the window. We settled down comfortably in our seats, while the flight attendants were keeping themselves busy with the needs of their passengers. Since it was a very full flight, they were scurrying around feverishly. I was reading and my mom was looking out of the window. Suddenly, one of the flight attendants stopped in her tracks, right at our seats and turned towards us. "Wow, the energy right here is fantastic. It is so vibrant!" she said. She looked directly into my face and asked, "Who are you? No, don't tell me yet. I'll come back, when I have more time."

My mother looked at me and exclaimed, "What did she say!" I repeated her words and my mom shook her head. "That is really something! You know I'm your mother, so of course, I am prejudiced. It's nice that someone else knows how special you are. I just shrugged my shoulders, as I still was shocked at the flight attendant's words.

When the attendant returned to our seats, she said that she was metaphysically attuned and sensitive to people's auric energy fields, especially related to frequency and vibration. In the frenzy of airline flights, most people are at their lowest ebb vibrationally. She said that she found that she got drained by those customers' energy fields and that it made her job more difficult. When she approached my seat, it was as if a wave of fresh, cool energy washed over her and it acted as a magnet. Her feeling was that it was highly unusual and that is why she wanted to know who I was.

When I protested that I was no one unusual, just an average cosmic being, she smiled and shook her head. "Oh no, you're not. It seems to me that you are special. Keep up the good work."

I can't say for sure what this means but if Zarg were here, I know he would say, "of course, you are, you're from Excelta. But like my mom, he, too, is prejudiced.

THE HELPING HAND

In 1974, my ex-husband and I were new citizens of South Florida. We moved our company's home office to what we felt was paradise. My parents also were new arrivals and lived close by. It was an exciting time for us, as our life was unfolding in a new way. I was focused on practicing being a wife, since I had only been married a few years. It was all a fresh start for me. I was not involved in anything metaphysical or paranormal in classes or reading. My spiritual life was still in its dormant stage.

My husband requested that I come to the office for some purpose which I do not remember. I had gone out to lunch with my mother, so she came along with me to see him. I was driving our car that day, with my mother seated next to me in the front seat. The office was situated near Federal Highway, a main traffic artery in Deerfield Beach. It is a divided road with two lanes of traffic going north and two going south. I wanted to cross the north lanes to get into the south lanes. It was near rush hour and the traffic was beginning to pick up. I carefully crossed the north lanes and was waiting in the median lane to pull into the southbound lanes. I waited until all of the traffic was cleared and then I turned south into the left lane. It was at that moment that a very strange and unforgettable event happened. As I made the turn and was headed south, I felt a hand pressed hard against my right shoulder, it felt as if it was coming from behind me. I looked up into my rear view mirror and saw a new black sedan almost flush against the back of my car! As soon as I looked, I heard a voice in my ear saying, "Don't worry Mia, you will be O.K. You will be protected." I was astonished. I seemed to be on automatic pilot. There was no squealing of the brakes from the car behind me. My heart was not even racing. I felt no adrenaline rush. My mother continued on chatting, as if nothing had happened and we continued on to our destination.

I never forgot that experience, as the moment itself felt like it was frozen in time. My mother is an excellent driver who notices everything going on in the road. Yet, she did not say a word about the possibly horrible near-collision. She seemed oblivious to it, even though her view of the black car would have been good since she could see it in the side rear view mirror. I didn't say anything to her after the event, since I was still feeling the hand on my shoulder and didn't want to alarm her. After I dropped her off at her home, I had the deep knowing that something momentous had happened to me. I felt that fate had truly stepped in and supported my aliveness. I did, indeed, feel protected. Whatever hand clasped my shoulder, with loving protection, did so at a moment that might have changed my life and my mother's forever.

I never gave any thought about Guardian Angels before but it was clear to me that some force for good was at work for me and the blessing was in the awareness of that.

"Mia, do you remember when we put our hand on your shoulder?"

– Zarg

"Many times it isn't a guardian angel on the job, it is your own higher self manifesting for your protection."

– Zarg

248

DOLPHIN SWIM/HUMAN WHIM

I have had three near drownings in this lifetime, and yet I have a strange kind of symbiosis with water. I love to see it and don't mind riding on it or above it, but being in it is not for me! When I was a kid I learned how to swim, and even got my Red Cross certificate, but something happened to me when I got to teen age. Whenever I got into deep water, panic would set in and I lost all touch of the reality of the swimming expertise I had. The whys and wherefors of this quirk of mine can be supposed from now until forever. I have my own theories that cannot be proven. It probably has something to do with past life experiences and/or a trip on board a space craft. It really doesn't matter, since I have had to learn how to deal with this phobia in the here and now. It has taken all of my inner resources to do it because, more than anything, I wanted to swim with dolphins!

The archetype and image of the dolphin has been imprinted deep into my psyche. Even prior to the coming of my New Age mentality, dolphins fascinated me. Over 30 years ago, I contributed to organizations that worked in the interests of wild dolphins. I never saw a dolphin first hand until I moved to Florida. As soon as I saw the six of them perform, my heart chakra vibrated to maximum intensity. I fell madly, passionately in love!

When I read about dolphin/human swims, I knew that it was something I had to do, despite my own deep water terrors. I knew that somehow it would manifest! I did do the swim, but it did not come easily. Just like life, the best laid plans oft do go astray!

I found the perfect dolphin swim. It was a three-day, intensive seminar in the Florida Keys. It would be facilitated by an Australian who channels dolphin information. Boy, that sounded interesting, and I would get to swim with dolphins in a setting where I could wear a life jacket. I made sure of that! I asked my friend Penny if she wanted to go. She most enthusiastically agreed. I was happy, since I knew that she would help me with my deep water anxiety and wouldn't let me drown! She also said that she would drive and I would not have to worry about driving down there alone. I felt so happy. I was being supported emotionally and could concentrate on quelling my fears. The most important thing was that I would be with my adored dolphins, one on one!

The day before we were to leave, Penny told me that she couldn't go. I was very shaken by this but determined, nonetheless. I had told anyone and everyone that would listen about the impending trip. I had my suitcase packed. I wanted to go, no matter what! I didn't want to act like a baby! I didn't want any of my insecurities to even enter my cerebral cortex. I was above the mental gremlins that were knocking at my scull with fears beyond terror. No, not me, I am a full-fledged adult and meditator. Like Columbus, nothing can stop me now! Good God, was I wrong!

I went to bed early, since I now had to drive down to the Florida Keys and didn't want to get into any rush hour madness. Everything was packed and neatly placed by the door. I set three alarm clocks, including hand-wound ones just in case there was a power outage. I was psyched! Thereby hangs a tale of incipient madness. I had never up until that night suffered from any kind of insomnia or sleep deprivation. No matter what sleeping situation I had encountered in my life and in my travels, sleep was never a problem! That was always absolutely true, until that night!

I laid in the bed in every conceivable sleeping position known to mankind, with my pillows piled either on or off the bed. I was stark awake, no matter what soothing thought or nighttime remedy I tried. All I could feel was my heart beating madly in my ears and reverberating out of my rib cage! I played the radio and TV and alternately turned them on and off! I played deep relaxation tapes. The more I did, the more panicked I became. The digital clock winked mercilessly! How was I going to manage the long drive and the pressure of the swims without sleep?!

I had never had *no* sleep! It was horrible, ironically like a living nightmare! I was totally freaked. I was having a hard time breathing. Was I having a heart attack or just hyperventilating? These thoughts made me feel even more intimidated by the unknown. The feeling of blood pulsing through my veins was terrifying! I now understood what insomniacs went through.. By the time my alarms went off at 6 AM., I had not slept *at all*. I felt totally out of control and in a trance at the same time. It was the scariest moment in my life. There is nothing worse to me than not being in control, and I was not! I tried to put off leaving for a few more hours, thinking that might alleviate the stress of the situation; but it didn't help.

I looked at those suitcases by the door and tried to rationalize my feelings. I said to myself that I was stronger than the situation and would go, nonetheless. In a daze, I took a shower, dressed and locked my apartment door. With suitcases in hand, I waited for the elevator to take me down to the garage from my twelfth floor apartment. Despite the handful of vitamins, I had gulped down, I still was groggy. The elevator door opened and I stepped in. Suddenly, almost as if a hurricane gust and a giant invisible hand had come from nowhere, I felt like I had been whooshed and pushed back out of the elevator and into my hallway again! The door closed and I stood astonished by what had just happened. "Oh, my God," I thought, "I am going stark, raving mad!" What had just happened was unintelligible to me. Who or what did that? Did *I* do that? My first thoughts were that I am now going so nutty that I probably will have agoraphobia and not be able to leave my high-rise apartment. I felt like Rapunzel in the castle, only there was no Prince Charming to take care of me. I sadly trudged back to my apartment. How could I possibly drive the three hours to the Keys?

How could I face my friends and all of those others I had boasted to about the dolphin swim? I was mortified. I just couldn't go. I couldn't impose on one of my friends to drive so far. How could I explain the night that just passed or the elevator that I got "tossed out" of? Nothing made sense to me, and I was missing an experience

I had so long dreamed about. Most of all, I was totally exhausted and couldn't sleep! I phoned the organizer of the weekend and told her that I wouldn't be coming, even though I had paid fully for the experience. She encouraged me to come later on in the day, but I didn't even know if I could get down the elevator! I very sorrowfully declined. Since I could at least keep my financial agreement, I told her that I would not ask for a refund. It wasn't her fault but whose fault was it?

In retrospect, I believe that the terror and trauma of swimming in the deep water and my phobia related to it took over even my most heartfelt intentions. Had Penny decided to go with me, I would have felt better. She didn't and I was all alone with my worst and most abiding fear. It was a catalyst for a physical and emotional overload for which I was not prepared. Nothing I had ever done could have prepared me for this episode. What followed was a year of "fear of the fear". There was no anticipating the sleepless nights that followed. There was no pattern to my insomnia; it just happened. The palpitations continued and I visited a few doctors to get some grounding in this new problem. I visited two sleep labs. One determined that I had sleep apnea, a real bad sleep problem that could lead to death! The other lab said I had no such thing! As this is now a few years past my first sleepless night, I feel that I have triumphed over these experiences. Whenever these nights of uninvited alertness occur, I know that I will not die of non sleep and inevitably, I will catch a few winks. The same horrible night has not happened again since. Thank God!

I was able to get back into the elevator again and I later found out why I wasn't able to take that ride down to the garage. The answer came in a most unexpected way. Through a mutual acquaintance, I met Johnny Galloway, via the mail. Johnny was a senior citizen who was said to be in contact with extraterrestrial intelligences since his childhood. He had a group that went out into the Brown Mountains of South Carolina to invite contact with space vehicles. I had heard he and his group were often successful. A week after the attempted dolphin swim debacle, Johnny phoned me. I had never spoken with him before. After the preliminary social exchanges, he told me that he had a message for me. He said that the reason I was not allowed to leave my home to go on the trip was because in my extreme state of agitation, I would have had an automobile accident. He said that the accident would have had dire consequences and I would have been physically incapacitated. That was not what was planned for me. I needed to be whole physically, so I was stopped from going by unseen forces. They had tried to tell me this telepathically but due to my extreme stubborn streak and willfulness, I was going to drive anyway. They had to do something radical to stop me, hence, the elevator experience. Jimmy told me to be happy about that, as it was my cosmic destiny at work! I was dumb-struck at his words. It truly amazed me how he knew this, because I had never mentioned a word of the experience to him or our mutual friend! I asked him how he got his information and all he would say was, "All is connected." I certainly felt much better with his information. I had not stopped flagellating myself for my childish fears and lack of control. Johnny poo-pooed this. He said that my fears were well founded because of past life experiences that were being re-experienced. The drowning feelings didn't all have to do with water. I had some falls that had also led to death and the feelings of drowning in deep water and falling were tantamount to similar physical reactions.

The adrenaline surge from fear can be as debilitating as it is energizing. This is especially evident when attempting to sleep while you are fearful and anxious. I was trying to be bigger than the situation but the fear overcame me. This was compounded by the adrenaline rush that made my heart beat like a bongo drum. Since I had no prior experience of that, I didn't know how to handle it. Johnny said that if this happened again, at least I would have a better understanding of the problem and would deal with it more successfully until it no longer posed a problem.

Whatever providence sent Johnny to me, I am grateful. It is true that on occasion, I do still have some kind of sleeplessness due to many factors. Yet, I have never since endured the trauma of that night of nights. Thanks to Johnny, I never will.

"Dolphins will leave the Earth when man learns to live together. They are kind and spiritual animals whose mission is humanity."
– Zarg

THE EYE OF THE DOLPHIN

Five years after my first attempt to swim with the dolphins, I made my second attempt. I felt frustrated at not meeting my goals, as they are life defining and empowering. No matter what comforting words Johnny Galloway had offered, *not* swimming with the dolphins felt like a loss for me. In the interim years, I visited all of the places I envisioned as important. One after the other, I was knocking them off my mental agenda, like bowling pins. I was waiting for the right venue to gather my strength for my next try.

I received a mailing about Dolphin Camp. It was being put together by a group from North Carolina. The swims with dolphins would be in the Florida Keys. The focus of these contacts would be on "interspecies communication." That sounded like it was for me; interspecies communication was perfectly aligned with my mentality and spirituality! "This is it," I thought. I contacted the group and asked about swimming proficiency needed and whether I would be allowed to wear a life jacket during the swims. I was assured that all levels of swimming were allowed and that life jackets were available and acceptable. There would be multiple swims at different facilities, as well as in the ocean with wild dolphins. I wasn't too keen on the ocean swim but the others sounded just perfect!

I signed up for the camp. It was to be for six days. We would stay at a luxury hotel in Marathon, in the Middle Keys. They even had dolphins at the hotel that were part of a research project for an organization from the Midwest. So it would be dolphins, dolphins, dolphins! That sounded like music to my ears. Underneath my excitement, I could feel the fear welling up in my throat like bile. Was I going to crap out again? I was determined not to. I would make things as easy as possible for myself. This time, I would not drive down there. I would fly, even if I had to charter a plane. If I, again, had an episode of sleeplessness, it would not be such a big deal. At least, I would get to the dolphins. I also would be very selective regarding who I told about this jaunt. No more shame for me, not that it felt like such a shame any more, but just in case! I felt ready. I had all contingencies planned for mentally. I was starting out with a clean slate and no erasers. Whatever happens is for the best, but I knew in my heart of hearts, I would go for the gold and get it!

I packed for the Dolphin Camp week. I even bought a new bathing suit, I was so sure I would use it. I found a flight that left Ft. Lauderdale and connected through Miami and then on to Marathon. When I landed at Marathon, I found out that three other women on that flight were also going to be at Dolphin Camp. They seemed nice so I started to feel reassured. When we arrived at the hotel, the whole group got together in one of the private hotel meeting rooms. There were 24 of us, 19 campers and five staff members. We introduced ourselves and gave our reasons and goals for coming. I felt it was truth telling time for me. All of the others felt at home in the water and had various reasons for attending. I was the only one who had any apprehensions or bad past related to doing this. I felt that I had to ask for the group's support in indulging my insecurities and paranoia. I told them how badly I felt about asking for this but I really needed it. What I got was unconditional support for whatever was ahead for me. The staff said that they would help as much as they possibly could. I was assured that everyone that swims with dolphins has his personal catharsis, no matter what preconceived ideas they came with. I certainly was beginning to breathe easier, knowing I was in such compatible company.

There are three different places to swim with captive dolphins in the Keys and we were to have multiple swims at two of them. I would eventually visit the third on my own. Each place has its own rules and philosophies on dolphin swims. In each case, I found that those who spoke with us had total love and reverence for the dolphins. The primary concern always was for what is best for the cetacean. I never felt that the animals were forced to do anything that they did not choose to do. I learned that you cannot force them to do anything. They are intelligent and have their own wills. Sometimes they choose to cooperate and other times they are in a snit or don't care to be social.

We did not swim with the dolphins the first day and a half. That was preparation time. The leadership of the camp did not take these swims lightly. We went through certain processes for trust, communication and fellowship. We were split into groups called pods. The pods became familial, in a good way, not a dysfunctional way. The term "pod" was highly appropriate, since that is what dolphin groups are called. The pod structure is the essence of dolphinhood. The bonds are so strong within the dolphin colony, that it becomes like one mind, a group mind. It is very rare to see a rogue dolphin, one who is apart from the group or prefers to be on his own. Dolphins thrive on each other. Their love and camaraderie becomes palpable, when you watch them interact within their group. That was the goal of the pod work for us and I was so very glad that was the plan. I would need every one in my pod to get me through my experience.

We started our swims at the Dolphins Plus facility. They had very stringent rules. You must swim with mask and fins, snorkel if you choose, but arms and hands must be at your sides at all times. They did not like people to swim with life jackets, but Pat, a Dolphin Camp staffer, demanded one for me and I got it. We had two octogenarian California women in our group, who had plastic knees and hips. We were told that dolphin's sonar can zero in on these human oddity parts, as soon as the human is near the water. Dolphins like to bounce their sonar off them and swim right over to the person, as soon as they enter the water. They find it fun and interesting. Usually clicks and whistles from the animals are heard. It proved to be 100% accurate, for both of these women. They were absolutely delighted, and as we were watching them. one of the women sang an aria from La Boheme to the fascinated dolphins. It was a great game for the dolphins, as it was for us imagining what they were saying to each other in dolphin language.

I felt very encouraged by this scenario and I also felt threatened. How could I let two eighty-plus year old ladies be so successful and me not? Pat encouraged me on and said that she would be nearby in the water. All eyes of my pod and the others in the camp were on me. I tightened my life jacket and slid into the 15 foot depths. I wished that I didn't know how deep it was but it was clearly marked on the sides of the pool like environment. I hadn't tightened my jacket enough and the water pushed the buoyant jacket up towards my throat. I panicked! I could feel the adrenaline rushing to my head and my extremities.

Pat, seeing that look in my face, immediately came closer. She tightened my jacket and suggested that we swim in tandem, she, floating belly down, in front, and me holding on to her waist like a caboose! Pat then asked me to observe the four dolphins that had surrounded me. I was too terrified to notice them before. Indeed, they were there keeping a respectful distance from me. I was finally with them. I knew how sensitive they were. They must have felt my fear and trepidation. I didn't want them to feel that. I only wanted them to know how much I had revered them for years!

Pat was chatting to me reassuringly. We were actually swimming! That is, she was swimming, I was holding on and doing a back kick. I finally began to relax and check out what going on around me. The others in the water were having a blast! One female dolphin was with us constantly. She must have thought that she was my nanny and custodian. I was so happy to have her undivided attention. I began sending mental messages to her. I was so thrilled to have taken this first step, and it was a giant step for me!

The next few days were filled with more processes and dolphin adventures. We swam at the Theater of the Sea twice. We had the most loving male dolphin companion, Perry. His female compadre decided that she didn't want to participate. She had become quite upset that one of the small docks had fallen into the water. It upset her particular need for conformity, as she didn't like things being changed in her environment. There was no way she could be induced to play with us. She would swim around us but liked acting the prima donna. The trainers all were concerned with her actions but felt that she had the right to do her own thing. That was when I truly understood how these marine mammals are allowed to be their own sentient selves.

Every trainer that we encountered was incredibly concerned with the dolphins well-being. That took precedence over everything. I had in my mind some question about the activity of dolphins for the benefit of humans. My thoughts are not so harsh now. I believe that there is a place for us to experience them in captivity. It appeared that Perry, especially, liked to play in the company of humans. The second time we swam with him, it seemed he knew us. As we were coming onto the dock stairs to swim, he swam into the shallow area and looked at us, as if in happy anticipation of our getting together. It was like he was saying, "Hey, it's about time you came back here!"

Perry made a big difference in my life. That is, Perry and my pod. They took over where Pat left off. They also ferried me around, Perry happily swimming around and under us. I started to feel comfortable enough to even swim by myself! Perry giving me the biggest gift of all, in allowing me to hold on to his dorsal fin and towing me throughout the water. He even did the rostrum push, by pushing me with his nose, while I floated on my back. If it wasn't for my wearing a floatation device, I could even imagine that I was a co-star of a dolphin show! After a couple of swims with Perry, I felt like a new woman! I certainly felt a lot better about myself.

The day before we left, everyone except me went snorkeling. I wanted to go the Dolphin Research Center. I felt the need to visit there. It was the only dolphin facility we had not visited. It is a very impressive facility, the most physically beautiful. It abuts the Gulf of Mexico with beautiful sea water lagoons with low fenced enclosures. They had a guided tour which I joined. The trainer imparted much information about dolphins which. I, after a week of total immersion, could have given myself! We walked around the center and were introduced to a few of the dolphins and given their histories and relationship to each other. The human mammals were much less animated than the marine mammals we met. I had learned that dolphins love sound and action. If you want to get their attention you must make noise and move around, not stand there like a tree stump. The group were living trees, except me! I started to talk to the dolphins out loud and clap my hands and do a little dance to make myself noticed. The human stumps looked askance at my antics but I did not care; I was in my element!

The group moved on to the next position but I stayed to get a closer look at Sinbad, the largest male in the pod. He also moved closer to me. It was at that moment that I had a pure cosmic experience, one which I will forever remember! I felt as if I had fallen into the eye of the dolphin. It was as if Sinbad blinked and sucked me into his consciousness.

Telepathically, I heard Sinbad say, "Dear Mia, we honor your presence here. We are humbled to know that you love us so much, that you were willing to face your greatest fear just to be with us. Whenever you are again at the doorway of fear, remember the eye of the dolphin and look deep within it and know that we are with you swimming at your side." Sinbad then turned his other eye to me and swam away. I stood there in tears, so touched by this moment of unexpected bliss! Now, I knew why I felt the need to visit this center on my own. The week of practice in "interspecies communication" paid off in more ways than I could have imagined. The eye of the dolphin is now ingrained deep within my psyche and is an inextricable part of me.

At our final session of Dolphin Camp, I shared my experience with Sinbad. I gave my heartfelt thanks to all of the campers and the staff who supported me through my fears until the exhilaration of my triumph over them. It was a very emotional experience for us all.

My triumph over my fears was not just mine. I came to represent each person's personal demons. They saw themselves in me. As I had moved through the process of my life that week, they were with me and with themselves through me. In the end, we met each other within the circle of the pod, the family of man joins the family of the sea. We embraced in joyous union.

"If every human could look into the eye of a dolphin, the whole world would be transformed."
- Mia Adams

"Why do you think having fun and being humorous is only for lower spiritual levels? My sense of humor is part of my natural being. Humor is a basic essence and nectar of life on Earth and all other levels."

– Zarg

COSMIC "CARMA"

It had been a very special wedding shower. Geri had given it for Auriol also known as Barbara Glass. Auriol and Geri were involved in a group that channeled information from the Pleiades, a star system in our universe. I was not a member of the group but got to know Auriol through a mutual friend.

Geri's home is quite unusual. It reflects her and her husband, Sam's interest in metaphysics. Unusual crystals were placed strategically throughout, and pyramids and other Egyptobilia were everywhere. Auriol was engaged to marry Juan, a somewhat older Chilean shaman, within a few weeks. It was Auriol's first marriage and not the average pairing. They were certain that hands in unseen realms orchestrated their union. Auriol told me that space beings she communicated with predicted the match. They certainly were good matchmakers!

The shower was special in that much love and terrific gifts were lavished on Auriol by her friends in the Pleiadian study group. I was quite touched by the genuine friendship that was shared. It was a unique experience. Even though they were not bonded genetically as family, they were bonded by friendship as family. It was an evening well spent.

My drive back home from Pompano Beach to Ft. Lauderdale was leisurely. I was still basking in the warm glow of Auriol's party. It was a perfect Florida December evening, cool and in the low 70's. I rolled the window down on the driver's side and began my drive back home. When I got to Lauderdale, I turned off the well trafficked Federal Highway and cruised down Bayview Drive. Bayview is a street that winds its two lanes through quiet, residential areas. Most of the homes are single story and expensive. That evening there were few cars on the street, as it was almost midnight. I love it best when I have the road to myself and this was one of those times.

I had driven about three miles on Bayview humming with the music on the car radio. I had just begun to accelerate and was about a half block past the stop light on 26th Street. The street ahead and behind me was empty. All of a sudden, there was a jolt and my windshield was totally covered with what looked like sleet or icy snow. It hit with tremendous impact as my seat belt jolted me back and I hit my seat with a loud slap. I couldn't see anything out of the front window. Luckily, I had presence of mind to put on my windshield wipers and it whisked whatever it was totally away instantly! My heart was beating furiously but I kept on going. I was afraid to stop. I thought if someone did this purposely to stop me, I certainly didn't want to give them the opportunity to do so. I looked out of my rear view and side view mirrors to see if I could see a vehicle or persons who could have done this but I saw nothing. The street and sidewalks were still empty. Sleet on a balmy, Florida December evening did not make sense!

Then I noticed that there were some scratches on my windshield that were not there before the impact of the sleet. What could it have been that could have scratched the windshield and look like sleet? Luckily, I was only five minutes from my home where I could inspect my car safely.

When I got into my garage and begun to check my car closely, I was even more baffled than before. Not only were there scratches on the windshield, some areas of my hood had paint eroded away down to the primer and the grill had what looked like crystals imbedded in between the grill work. The grill also had pitted areas. Whatever had hit the car, hit it at an angle on the driver's side as my roof and side also had scratched and eroded areas. I had to be grateful though. I did have the driver's window open fully, and even when the full impact was felt, nothing came into the car. With such a corrosive material, I could have been hurt!

Since I had a Cadillac Eldorado, I knew this was going to be an expensive repair job. The big question was how was I going to explain this happening to my insurance agent when I couldn't explain it to myself? Everything always seems to happen on Fridays, so I had to wait until Monday to seek some answers. I went to the body shop at my local Cadillac dealer. I was hoping that they, with their expertise, could give me some idea as to the material that caused the direct hit. All I got were shoulder shrugs, shaking heads and narrowed eyes. A hypothetical guess that was totally eliminated was that it could have been frozen blue water accidentally discharged from a passing jet plane's toilets. There was nothing in the residue on my car that melted or was blue. It had the initial appearance and consistency of icy sleet, but it wasn't.

They asked *me* what it was that hit the car. The consensus was that they had never seen the likes of that before. Even the crystallization on the grill was strange. It looked like glass shards but disintegrated on touch and became powdery. I finally was able to get an appointment on the following Friday with the Allstate claims agent, he threw up his hands in futility and went into his office to get a camera. He said that he couldn't explain it but would take a picture of it. Since one picture is worth a thousand words, that would suffice for an explanation. He also gave me carte blanche on repairs, which made me happy, as I needed repainting on half of my car! No one seemed to have seen such strange kind of destruction on the exterior of a car before!

Even more fascinating than the lack of explanation, was the explanation. In my subsequent conversation with Zarg, he unhesitatingly knew the answer. He said that the stuff that hit my car was COSMIC WASTE, ASTRO DUST from a passing Pleiadian space ship! "Good God," I said, "why?" It seemed that they just wanted me to know they were there, kind of like a space hello, how ya doing, kiddo! Probably, they knew that I was with their friends at Auriol's party and they wanted to acknowledge me.

"Well," I responded to Zarg, "if they wanted to greet me in such a forceful, destructive way, they should also send down the $1200 it took to repair my car!" Zarg, being his old straight talking self, said, "Now you know you only paid $200 with your deductible." I guess you can't put anything over on those ETs!

*"You will lead the group
understanding in Peru about
what you will see.
We will take care of you under
the light and watch you on our
projection screen."*
– Zarg

ONE NIGHT'S LIGHTS IN PERU

I knew that I would find the right group to go to Peru with and my patience was rewarded. Oliver Mendez Quiroga was a true Peruvian-American. His father is a well known physician in Peru and his mother is American. He was raised in both cultures and utilizes the metaphysical component of Peruvian culture in his work as a therapist. He also was connected with a Peruvian contactee group that is well known for its abilities to mentally contact and bring in space craft. His connections with like-minded Peruvians was a perfect templet for the trip's focus. I won't dwell on the virtues of seeing Peru, especially Machu Pichu and the Nazca Plains, for every human, seeker and non-seeker alike. Peru's reputation is well known to all travelers. I did have one special night that stands out.

We had spent the day in the colorful and bargain-producing market town of Pisac. Zarg calls me the "Shopping Queen," and I sure earned the title that day! When we left Pisac, we headed for the Spanish colonial looking town of Huambutio, not too far away. We stayed in a very rustic hotel that was a former retreat for Catholic clergymen. There was no central heating for the cold nights but each bed held at least one llama or vicuna pelt blanket to keep us toasty.

Huambutio is known for its magnificent star and space craft gazing. The town is surrounded by mountains and dense lush vegetation. On a clear night, stars fill the skies with views that are unique to that hemisphere. It had never occurred to me that the constellations are different there. My favorite was the Magellanic Cloud which is only viewed from the southern hemisphere. It is either of two irregular galactic clusters that are the nearest independent star system to our own. Even in a dark night sky, it will appear as a giant cloud, not as separate stars. I was amazed when I first saw it.

We had finished dinner and were sitting at a long refectory table. We were passing the time until it was dark enough for us to go out and look at the sky, with the hopes that we would see space craft. The area around Huambutio has a legendary history of sightings. Many of the locals feel that there is a base for space craft in the mountains there.

A couple from Atlanta had brought rune stones and were "throwing them." Rune stones have characters from ancient languages, especially Scandinavian. Each character is symbolic and can be interpreted by those who are savvy in this knowledge. They were giving us each a turn to pick a stone and then they would analyze it. I was only vaguely aware of this clairvoyant or psychic technique. When my turn came, I picked a stone with five dots on it. The way I had it oriented was with three dots on top and two below. The couple's explanation of my rune was meaningless to me, it meant nothing in relationship to my life. I shrugged and thanked them anyway. Finally, Oliver told us that it was now dark enough to do stargazing and said that since it was getting quite cold outside, we should get our warmest clothes on and bring down our pelt blankets to cover us. My roommate, Melanie, and I trudged upstairs to our room and we layered ourselves with as much clothing as our South Florida constitutions warranted. We grabbed our blankets and started to walk out of the room. Melanie was very excited. She was new to things ufological and our discussions left her wide-eyed in amazement.

As I was closing the door of our room, I got an intuitive feeling that came as a voice in my mind. I hesitated to mention it to Melanie because I did not want to disillusion her, if I was wrong. I said to Melanie, "I know what my runestone means, it wasn't what they told me. I have a hunch that we will see some UFOs tonight and they will be positioned in the sky like my rune stone. There will be three ships above and two ships below them. I think that the larger ones will be above." Melanie gave me a big smile and said, "Boy, I sure hope so, that would be a dream come true!" I agreed with her and we walked downstairs to the garden terrace where our group was gathered. As we approached the garden door, Oliver was standing there. I decided to tell Oliver what I had heard telepathically. When I did, he smiled at me, with a twinkle in his eye. "Yep, you are so very right!" he said. "They are out there right now, go and see them, they are straight ahead, right above the horizon!" As we went out, we looked in the direction Oliver stated and sure enough there standing out in a star laden night sky were five bright pulsating white lights, three bigger than the other two. They looked much closer than any of the stars and appeared to be vibrating. Most of the group was there and were also looking with amazement at the lights.

Melanie was jumping up and down. I was very happy that I had told her, despite my ego's fear of being wrong. We watched for approximately two minutes and then the five bright lights blinked out together and disappeared. I had a feeling that they were there for me. It was another validation of my ability to connect to the Universal Mind, when I allow it. How the runestones were manipulated, I can only guess. Did I create the stone symbol for myself, did those on the unseen level do it, or was it another example of the innumerable "coincidences" in my life?!

We watched the sky for an hour or so, until we were turning blue with cold. We saw some movement in the heavens but nothing we could identify as anything really unusual. It was a night to be remembered for me, when I sincerely felt that the universe and I were one!

"Put behind the robes of the past. Out of chaos comes order. It is demonstrated in the Mandelbrot set in the English fields."
– Zarg

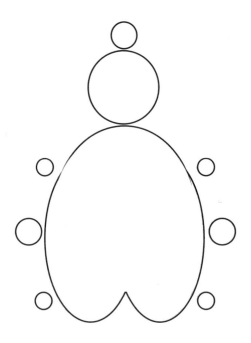

CIRCULAR VISION - ENGLISH STYLE

In 1990, I had been a UFO researcher for less than two years, when I made my vision quest to England. A local New Age store sponsored a trip to England. I was the only one of the dozen of us wayfarers who knew anything about UFOs or extraterrestrials. This was not my first trip to England, as I had been there several times before. I had even lived there for a year, but it was a different view of England that I was now seeking. I had not been metaphysically oriented during my other trips. On those trips, I was functioning solely on the touristic level. When I lived in London, I was quite young and it was in the heyday of the Beatles, when England was the center of all that was hip. My focus then was strictly on fun and games. It is interesting how far I have come and how life generates so many different roads to travel. The path I hoped to trod this time around was one centered on spirit and communion with extraterrestrial sources.

This period in England was pivotal in the exploration of the crop circle phenomena. The circles had begun to proliferate and were first being taken seriously in the late 1980's. Those, who were investigating circle sites and the scientific evidence involved in the investigation were now called "cereologists." The circles were found mostly in cereal crops, hence the term.

I was very fascinated with the circles and their often incredibly beautiful designs. Despite some hoaxing of circles, there were others whose mysterious appearances could not be denied. I just knew that there was something beyond obvious cliche scientific explanations of these artful and powerful designs in the crops. I knew that unseen forces, spiritual, interdimensional or extraterrestrial had invisible hands in their creations. It was these beauties that called to me. Like the Sirens of the deep, beckoning to sailors, I just had to be with them, well, at least one! On the flight over the Atlantic, I snuggled into my seat, closed my eyes and envisioned myself in a crop circle.

I smelled the freshness of the crop as I lay mid center in the formation. I saw myself fully open my arms and move them away from my sides, moving them up and down, making "angel wings" in the crops, just like I did in my childhood days in Chicago's snow! I had to experience a crop circle on this trip! It was my raison d'etre for coming to England, this time around, and I knew I would. I had no doubt of that!

We were met at Heathrow Airport by Virginia, who owned the metaphysical center we were staying at. Virginia is a spiritualist minister who has a very large, lovely home in Shepton Mallet, Somerset, called Chalice Haven. Virginia opens her home to pilgrims on the path to the wonders of south central England. Many splendored wonders to be sure are found in the neighboring towns of Glastonbury, Bath, Avebury, Cheddar and Wells. Stonehenge was also close by and was our first stop, directly from the airport. Synchronistically, this area is also what the English call the "UFO Center of the World." Many close encounters have been written about in this part of England. Unexplained lights in the sky were normal here. I knew in my gut that I was in my rightful place and with the right group. I did have a lot of educating to do. Since this group were neophytes in all things ufological and knew nothing about crop circles, I had my work cut out for me. They were willing students though.

I tried to press the issue of visiting or finding crop circles with Virginia but she would not have it. She just shoulder shrugged my inquisitions and directed my inquiries to certain geographical landmarks, like Warminster, Silbury Hill and Cley Hill. For some reason, she just didn't find my focus that interesting. She felt that our limited time there would be better served visiting other venues. Warminster is a nearby military town where UFOs have historically flown over regularly.

Cley Hill is a man made neolithic mound that was quite close to Chalice Haven. We had driven by it several times. Virginia noted that UFOs were often seen over the hill and visitors often camped out over night, just to hopefully get a glimpse of something otherworldly. I was chomping at the bit to visit and climb the hill myself. I had a good feeling about that site.

Virginia and her son in law, Geoffrey, alternately drove the van that carried us to our destinations. Geoff was a real character, I liked him a lot. He was the antithesis to the stuffy Britisher; he was a down home type of guy who loved a good joke. He had a lot of combustive energy that was well tamed by his wife, Constance and their adorable tyke David.

We were into the second and final week of our trip and even though we had some great and unique experiences, I still hadn't done my angel wings in the crops! I was beginning to get depressed about the prospect of not experiencing my heart felt vision. I still had hope, we had three more days before we left the area.

Virginia drove us to the ancient Roman town of Bath. Despite the remnants of the impact of the Romans on the town, there was also evidence of wonderfully modern shopping in the enclosed malls, flea markets and closed to traffic shopping streets. After an informative tour of the Roman baths of Bath, we had free time to browse and carouse. I took off on my own to a unique shopping street built on an open bridge. Having been an entrepreneur and holding a black belt in shopping, there is nothing I prefer better than browsing. I looked into a jewelry shop window that had some very interesting unusually designed silver earrings. I walked in to inquire about the price. The shop was busy and I looked through the display cases, while I waited. Amongst the new and shiny creations, I spotted what looked like 2 very ancient rings. They were quite delicate and tarnished to a very dark patina. They looked like museum pieces to me. Often in museums, I worked my way to the displays of jewelry from the distant past. My mind fantasized about the owners of these little beauties. One of the rings had a small clear, blue green stone with a cross etched into the top, and if you squinted you could see it very well. The other had a flat rectangular top sans stone. I looked around the other showcases but I did not see any other ancient or antique pieces.

When the salesman was available, I asked about these rings and if they were for sale. "Indeed," he replied, "they are quite old and very unusual finds, especially the one with the glass stone. It is an original and intact Roman finger ring, inset with a clear glass stone from the third to fourth century A.D. It is very rare in this state and incredibly, it is still wearable."

"Wow," I thought, "I must have this. I always wanted a museum type piece of jewelry. It must be very expensive, I hope I can afford it." Miraculously, I could! The price was only $100. Someone had found the rings near a Roman excavation site and sold it to the shop owner. Since he wanted to make a quick sale, he made the price reasonable. Part of the bezel holding the stone was gone but the ring was still strong, so it was a bit less than museum quality. It was quality enough for me.

I got a receipt authenticating the age and origin of the ring. As I put it on my finger, I had a strong feeling that I had worn this before. Somehow I felt that this serendipitous find was not so serendipitous, that it was mine by the divine right of consciousness. I got the feeling that I had once lived in this part of England. I knew that this ring had been mine, was lost and again returned to me. Now that the ring was on my finger, the energy was palpable. I had always felt a close tie with England and things English. My bond to that country was strong. It always felt like I was coming home, whenever I arrived on the soil of England; now I knew why.

I happily left the jewelry shop thinking that this purchase was a good sign. I just got one of my heart's desires and now experiencing a crop circle was inevitable; it *would* happen.

I moved on to the flea market area and spotted the used book stalls.

Finding British books on UFOs is very difficult in the United States, especially older writings. I went from stall to stall asking for them. At one stall, the proprietress said, "You're the dozenth person to ask for these books today, it must have been that telly program last night that's created this interest."

"What program was that?" I inquired. "The BBC had a special program about a new crop formation that is quite near here. It is the largest and most complicated one yet. They moved their cameras and all of the electronic equipment into the circle and not only did the equipment fail, but it also blew up! They had to broadcast the program from outside of the formation before anything would work."

Now that was what I call synchronicity! Talk about the divine order of things! I knew that the ring would bring me good luck. I felt like I was being challenged to listen to my inner guidance. It was like being Alice in Wonderland; would I find the right doorway to pass through to get to my heart's desires?

I asked the book stall's proprietress if they gave out the exact location of the crop formation. She stated that they are always non-specific about the locations, since the farmers don't want people tramping about in their fresh crops. The area was only generally given as Wiltshire, nothing more.

I happily returned to the waiting van and my fellow travellers. I showed them my ring and told them the news of the new formation. We discussed the possibility of finding the site. When we were almost back to Chalice Haven, we were passing Cley Hill, the UFO sighting spot. I asked Virginia if she would stop so that I could climb to the top and affirm via meditation that we would all experience this new crop formation. The group all enthusiastically agreed to this departure from our day's itinerary. Luckily, the day was unusually clear and balmy and perfect for the short climb to the top. I was joined by the two men from the group, while the others waited patiently in the van. As we were climbing and were gaining some altitude, we looked down into the field below. Quite clearly, we saw twenty little "grapeshot" type crop circles. These are small, clear circular imprints with only one ring. Some were close together and others further apart. The three of us were quite delighted with the view of these. We meditated for about ten minutes and found our way back down to the van.

Soon after we arrived at Chalice Haven, Geoff appeared. "Mia," he said, "I saw this interesting program on the telly last night about a new crop formation."

"I know about the program," I replied, "do you think that we could find it?"

"I rang the television station today to see if I could get some particulars about the location, but they said that they had to keep it secret. They would only tell me that it was in Wiltshire. Wiltshire is a large area. We would have to know more specifically where, to be helpful," he responded.

"I have got to get there, if I don't I will be so disappointed. We just meditated on Cley Hill, so I know that it is possible." Geoff thought for a bit and then got a very mischievous grin. "Yeah, we will find it. I will make this my personal project. I am off from work today and tomorrow, so I have the time to figure this mystery out. I can't let you Yanks think that you and the Japanese are the only ones who can get a job done. I have an idea how to find out where this circle is. First, I'll ring the station again and ask for the bookeeper. I'll pretend that I am the van driver that took the BBC crew to the site, the other day. I figure that they would have to refill the van with petrol and probably did it in the general vicinity of the site. I'm quite good at chatting up people, I'll tell the bookeeper that I forgot to put the location into my log. Then once I know what town it is near, I will call the local police. They know everything about what is going on. If I don't get any satisfaction with the police, I will call the local pubs. There is no way that those folks who hang out there don't know what is going on and they won't be so closed mouthed. Have faith, Mia, you'll find your crop formation!"

I hugged Geoff. He was the right one to get the job done. His mind could work wonders, I had no doubt of that. He went to the phone to begin the process. Listening to him "chat up" the female bookkeeper at the television station made me and the others' eaves dropping smile. He was smooth and got results! Terminating his conversation with the bookkeeper, he gave us the thumbs up sign. Early victory was ours. Geoff got the general vicinity in Wiltshire where they gassed up the van. He then called the closest police station and got some additional information to phone another town's police. When he did, the response was closed mouths, so that in itself was an answer. He then went to the phone directory and found some pubs in that town. When he called them, he received the answer he sought.

Since the crop circle was in an uncut field, we needed specific directions. We got the names of intersecting roads, so we were now in the ball game and ready for a home run! We decided that Geoff would drive us the next day to the town of Devises, where the crop circle was supposed to be. I was so excited and my enthusiasm was infectious. The others in the group were happy to forgo our final day's visit to local castles, just to see me get my vision made manifest and they also to get this unexpected experience with me.

Geoff arrived right on time, beaming and proud of his taking firm hold of the helm. He had never been to Devises before but knew how to find the town. He was told that the field was near the local golf course, which was a perfect landmark to find. Eventually, we found the golf course, but the area was extremely flat. Since the fields were deep in fresh crops, it was hard to see the crop circle from the road. One needed a vantage point that was elevated to see the form of the circle.

This was the first point of frustration for us. We drove around for a half hour. We knew that we were close and even after getting specific directions, it was hard to find the exact spot on the road to stop. Finally, Geoff saw a man walking down the road. He got out of the car and asked him if he knew where the crop formation was. The man took a moment before he answered, eyeing Geoff and us in the car. "Yes, I do. It is in my field," he replied. Geoff, not missing a beat, asked if we could have permission to enter his field, knowing that it is illegal trespass without permission. The farmer eyed us again.

Geoff's gift of gab was not to be unused. "Sir, we would be so appreciative if you would allow us to. There is a famous UFO researcher from the States with me that has come here just to see the circle in your field. She would be so disappointed if she couldn't. Please allow us the privilege of entry."

How could the farmer say no? All of our eager and expectant faces fixed on him and his response. Silently he nodded assent and pointed to the field right next to where the van was parked. He said that the circle was inside about 50 feet from the fence and showed us how to get in so that we wouldn't trample the crops. We were all jubilant! The Universe was in full compliance with us again! Of all the people on the road to ask, we picked the right one and now even had permission to go in! The crops were so tall and thick that it would have been nearly impossible to have found this on our own, without specific direction.

We all piled out from the car and made way towards the circle. I was in Earthly heaven! Once we had orientation within the field, we found the circle. It was huge. The circle, comprised of three other concentric circles within, had a total diameter of 140 feet. It was the largest of its kind that year. Since then there have been many others that have surpassed the size and have had far more intricate designs, but this crop formation was unique for its time. We walked and played in the formation, as if we were kids. We had come prepared with high, waterproof boots, so we reveled in the crops. I quickly found my way to the center and laid down. Just like my vision on the airplane, I closed my eyes and smelled the pungent odor of earth and wheat. It was delicious and I was exhilarated. I separated my legs and my arms from my body. I moved my arms up and down and made "angel wings." I prayed thankfully to God and all of the Powers that Be for getting me to that precious moment in time. Lying there, I went over all of the preliminaries that got me there. They all worked for good. I could feel the hairs raise on my arms and the back of my neck. The electromagnetic effect of the crop formation that undid the electronics of the television equipment was obvious. The others were also remarking on the tingly feelings they were having. We did a short meditation within the circle and after a half hour we left.

I thanked Geoff. He had made my dream vision come true; maybe we all did it together. Perhaps all of our energies created the desired results. The others thanked me for being the catalyst for their experience. None of them had been aware of the crop circle phenomenon before and now they had become front row participants to it. We all couldn't stop smiling on our way back to Chalice Haven.

The experience in England that year was pivotal for me. I had the sense that from then on, each experience was going to be sacred and an integral part of my development spiritually. I knew the power in vision quests for soul's growth and evolvement. I also had the sense that time was speeding up. It was incumbent upon me to accelerate these experiences to facilitate my moving on to my next level of awareness.

Upon returning home, I had a chat with Zarg about my trip to England.

MIA: Are you aware of my experience in the crop circle?
ZARG: YES, WE SAW YOU ON YOUR BACK, WE

LAUGHED. YOU HAVE ACHIEVED GREAT CONTACT WITH THE FORCES OF NATURE AND THE COSMOS.

MIA: Did you help me find them?

ZARG: THEY CAME TO YOU, YOU HAD TO BE THERE.

MIA: Was Geoffrey there to help me find it?

ZARG: YOU BROUGHT GEOFFREY INTO YOUR EXPERIENCE WITH YOUR ENERGY. YOU ARE VERY POWERFUL.

MIA: Does Geoff have Karma in common with me?

ZARG: NO, YOU ARE A MASTER, YOU ARE GROWING UP BUT YOU MUST BE VERY CAREFUL AS YOU HAVE GREAT POWERS NOW.

MIA: Psychic powers?

ZARG: SPECIAL KINETIC POWERS, ENERGY, WITH PEOPLE WHO WILL LISTEN, HERE AND NOW AND IN THE FUTURE. YOU ALSO MUST BE PREPARED TO HELP OTHERS INTELLECTUALLY, TO UNDERSTAND THE PROCESS OF LIFE ON EARTH AND IN THEIR COSMIC FUTURE.

COSMIC ENERSPHERE

Nina Munoz, Claire Marvin and I were looking forward to our second trip to Puerto Rico to do UFO research in May, 1992. We were accompanied by Sheryl Peters of Maine, a UFO researcher, who was making her first journey there. Our trip the year before was not only a lot of fun but was a real learning experience. *(See "SPACE BASE PUERTO RICO", page 45)*

An outstanding occurrence happened towards the end of our second trip. The year before, Nina had mentioned that there was a pyramid built in the backyard of a private home in Moca, a town in the northwestern part of the island. It is a very scenic area of lush, rolling hills and lovely homes. The pyramid had been built by a woman who was an abductee/contactee of extraterrestrials, probably Pleiadians. I say Pleiadians due to information I have received. Unfortunately, I did not have time to directly question her so this is what I was told. We had heard that the pyramid was built with the same specifications as the capstone of the Great Pyramid Cheops of Egypt. The Moca pyramid is now used as a meditation and healing chamber. There has been much documentation of healings, spontaneous and other that have taken place inside the pyramid chamber. often, disembodied hands had been seen moving over the healing subject.

I was determined to go to the Moca Pyramid on this trip. I had made a trip to Egypt in January of 1992 and had the incredible experience of being able to meditate with a group inside the King's Chamber of Cheops for three hours one evening. If I could experience the Moca Pyramid and meditate within it, I would have been inside the complete Cheops, since it was as if I had been experiencing the incredible energy of the Cheops capstone also; and I would be doing this all within four months! What an extraordinary miracle that would be for me. I had to go!

I convinced Nina, Claire and Sheryl without much effort and we went to Moca, accompanied by Alan Rivera, a well-known abductee who had been the subject of much media coverage. The way to gain entree to the pyramid is by going and asking at the owner's door! Apparently appointments are not taken, it is just a matter of luck, timing or the evaluation of the visitor's energy pattern by the owner and/or the E.T.s in residence there.

"Megin," the owner, is a full-bodied woman who is a mother and grandmother. Her home is large and ranch style with a beautiful, colorful garden. The day we visited her was an important one for her family, since a grandchild, just a day old, had arrived home. Apparently, this was an extended family with several generations living together under one roof. It was not the best day to impose on the privacy of the family. But, the fates were with us! Megin recognized Alan from his public appearances and remembered Nina from a visit she had made there the year before with a friend who was originally from Moca and knew Megin well. At first, she was reluctant to allow us into the chamber, since she was "cleansing" it from energies of other pilgrims who had come to experience it. That took several days to complete. With some polite begging on our part, she kindly allowed us to go into her backyard and into the pyramid.

Since we did not want to take undue privilege of the visit, timewise, I was not able to measure the exact size of the pyramid. It was large enough to have a dozen people inside and appeared to be made of cement painted white. There was an antechamber at the entrance, so that you can take your shoes off and leave them there, prior to entering. The floor was painted cement with throw-carpets positioned around the chamber. In midcenter, there was a large crystal imbedded into the floor. The smell of incense had permeated the chamber, even though there was none lit. Since it was a very humid day, the chamber was stuffy. Despite that, the feeling of a powerful energy was palpable. It made my ears ring and my head feel full. We sat down and I did a short meditation. Soon after, Nina felt she had to leave, since the chamber felt too energized for her comfort. I was the last one to leave. As I put on my shoes in the antechamber, I thought that I should take a photo of the altar that was straight out ahead of me about 12 feet. At the same time, I had the "thought," Claire came back to ask me to take the same photo. I told her that I was going to. We had been told *not* to take photos INSIDE the pyramid. I did not question why. I am sure that Megin had her reasons. Pictures outside were permitted and I took three.

It has often been said, "there are no accidents in the Universe." I think that this maxim holds true for the photo event that took place. I have the feeling that Claire also picked up the mental suggestion for me to take the photo of the altar. The few seconds that I trained the camera on the altar has made a difference in the lives of those of us who were here. We had a feeling that the Moca Pyramid was a special place but now we *knew* that it was!

The altar, directly facing the entrance to the antechamber, was approximately six and one half feet high by four feet wide. It was composed of large, rough boulders and stones. Interspersed amongst those were various sizes of crystals, amethyst and other semi-precious stones, and two statues of the Virgin. Surrounding the structure were abundant trees and bushes with climbing foliage around and over the rocks. I thought it would, indeed, be a very pretty picture for my memory bank and my photo album. It proved to be much more.

When I had my roll of 200 ASA, 110 film developed, I was in for a major surprise! The shot of the altar had a very large, transparent sphere sitting or moving smack dab in the center of the top section of the altar! The lower center of the sphere was contoured exactly the same as the rocks below it, as if it was making an impression on the form. On the periphery of the near perfect sphere appeared to be vestiges radiating out from it. The sphere itself appeared to be bluish-white against the background of the altar and the greenery, but everything behind it was visible, although seen in the same color of bluish-white. What could that "thing" be? My first thought was that there was some imperfection of the film, or that some kind of circular sticker had gotten on the film, or something was on my lens or something else rational. It wasn't a lens flare, since the bottom followed and curved around the rocks' edges. I had taken a photo of the pyramid before the altar and after so I thought that I could use them as a kind of "control." Both of them were totally normal, although Nina showed me what she thought was a similar spherical shape up in the trees in the photo of the pyramid and the altar from afar. When I checked the altar negative, I saw that the spherical shape was a dark spot on the negative so whatever I photographed *was*, indeed, on the negative.

After showing the photo to my friends and several photographic professionals, it appears that I had truly photographed *something*. It is something that is in another aspect of dimension, not in our third dimension, or we would all have seen it! It has been analyzed as moving, probably slowly over or in front of the altar. It is reasonable to assume that it wanted to be seen and photographed by me. Claire sensed it and so did I. I used a simple, old Minolta camera. It was not special equipment. Since this pyramid has been said to have alien energies working for and with the human beings who come to be healed or spiritually uplifted, I believe that this kind of energy was what I photographed.

In the history of ufology, beings and ships have been seen to change shape from amorphous and interdimensional to substance. This photo is an example of the same. I have been given the privilege of photographing this thing. Whether it is a being or a ship or probe, I cannot say for sure. My hunch is that it is a being, sentient and physical that we in our own third dimensional bodies and landscape cannot see with our physical eyes. It does exist, nonetheless.

I have given this subject matter a name. I call it a Cosmic Enersphere, for that is what it is; a sphere of cosmic energy. The Universe had worked in a magical and exciting fashion for me again! To Infinite, Divine, Creative Intelligence, I give a huge thank you!

Cosmic Enersphere ©1995

Eventually, those that knew Elenita were sure that she was the incarnation of the Virgin. She emulated all of the traits of the Holy Mother, hence, it was not an unusual assumption. Elenita certainly wasn't your average 19th century woman.

THE LADY OF THE SACRED MOUNTAIN

The reported occurrences at Santa Montana, or the Sacred Mountain, exemplify the level of high paranormal strangeness that has always existed in Puerto Rico. Santa Montana is located near San Lorenzo, a small town south east of Caguas. It is a quite rural area and was even more so in the late 1890's. One day, seemingly out of nowhere, Elenita appeared. She was not from a local family. Where she came from and who her parents were is still a mystery.

Elenita did not look like the local people. She was average height but very pale and extremely frail. Her hair was light brown and her eyes were greenish hazel. Her clothes were not rustic like the others. She wore a long, flowing gown of a light silky material. Most of the time when she was in San Lorenzo, she was dressed in a garment similar to the habit of the Carmelite nuns. When she thought she was in complete privacy, she had been seen wearing a crown and assorted fabulous jewels. She was like a vision.

It seemed that she glowed with sweet and gentle love that was also powerfully vibrant. Her ways were also different. Initially, she lived under a huge boulder, which she was able to lift or levitate at will. Many men together could not move this object but wispy Elenita did. The townspeople eventually built her a small home where she lived until she died. She was a vegetarian who ate selectively. Mostly she was seen eating sour lemons that no one else ate, but when she offered them to the locals, they said they tasted sweet.

Although she had great wisdom in many areas, she kept her physical distance from people. She would only allow, young pre-pubescent girls to touch her. Once they sexually matured, they could no longer act as her disciples, servants or helpmates. Elenita encouraged spiritual discipline and universal love. She was well-versed in Roman Catholic matters and encouraged the rituals of the Church. She was especially interested in interpreting what the Bible meant about Lucifer. She said that Lucifer had fallen from God's grace and came not only to Earth but went to different planets and took different forms. Could she have been giving the Puerto Ricans an identity for the extraterrestrials that inhabited their island who were negative and whose presence would be so pervasive later on in the twentieth century?

Eventually, those that knew Elenita were sure that she was the incarnation of the Virgin. She emulated all of the traits of the Holy Mother, hence, it was not an unusual assumption. Elenita certainly wasn't your average 19th century woman.

She made many predictions about the 20th century that have since proven correct. Many of them have materialized as our political history has unfolded. She also foretold precisely of our great advances in technology. There are a few which have not yet come to fruition, but many have. Elenita said that Puerto Rico would have great cataclysmic events happening towards the end of the 20th century. The island would mostly be under water as a result of a huge earthquake emanating from the ocean originating southeast of the town of Yabucoa. It would effect the coastal areas of the island greatly. Most would be destroyed. She stated that "at the time women will wear pants, the floods will occur". She went on to say that the area of Santa Montana will become the center of Puerto Rico and it would be the seat of power. The name of the town will change to "The New Aurora." Men will no longer head the government and machissimo will no longer prevail. Women will be given the chance to be as powerful as men had been, and will be at the helm. Elenita advised that continual meditation and prayer will be needed to get through those tempestuous times.

Not only was Elenita able to show exceptional physical abilities related to levitation, she also showed telepathic abilities by being able to read people's thoughts, even those who lived far away and whom she had never met. In addition, she was able to control nature by making crickets stop chirping and transforming herself physically into a dove. These abilities were considered totally miraculous to the humble folks in the village. She spilled her blood during Easter, saying that wherever her blood was spilled would be sacred ground.

Elenita lived happily in San Lorenzo and on Santa Montana for less than 10 years, until she died peacefully. The townspeople pondered what to do with the body of someone who was a saint or at least saintly.

She had given explicit information that upon her death, her body was not to be touched for three days, and specified that a glass sarcophagus be created for her. Upon her death, several men, whom no one had ever seen before, claimed that they were Elenita's family and wanted to take her body. When they started to carry her sarcophagus down the mountain, a fog started to form. By the time they arrived at the bottom, the fog became dense, and when the townspeople peered into the glass sarcophagus, Elenita's body had disappeared.

Who Elenita truly was remains a mystery. Because she lived there, Santa Montana has become a site of pilgrimage. A Roman Catholic chapel and religious book store are adjacent to Elenita's home and the places she meditated. Many strange and paranormal events have occurred there in the past and continue in the present. Reports of extraterrestrial sightings, UFOs and abductions are common. At one religious ceremony, where hundreds of people gathered, it is said that an alien was seen trying to take a small child away. The child was rescued by some men, in a tug-of-war! Currently, NASA has shown interest in Santa Montana. The chapel site is now closed at sundown. It is forbidden to go there at night, unless you are connected with the U.S. government or NASA. Many people have reported seeing marked vehicles and individuals with uniforms bearing the NASA logo. What could possibly relate to this show of interest? Perhaps, I have an inkling of fact related to this.

Thanks to my wonderful little Minolta, another photo insight has happened. When I visited the mountain with my friends, I had a very intense urge to aim my camera over the huge crucifix at the top of the mountain. It felt a little strange there and I had the desire to photograph the overcast cloud formations surrounding and especially above the giant crucifix. When the photo was developed, it showed a brilliant blue green beam shooting down out of the clouds.

After we left Santa Montana, and prior to my developing the photos, when we told Juan Marron, the Puerto Rican UFO researcher, that we had visited Santa Montana, he told us many interesting facts about the place. Included was information about the giant crucifix I had photographed. Juan said that the crucifix appears to be the entry point, or dimensional doorway, for space craft. He said that often blue beams and ships are seen coming right out of the sky, over the cross! It seems that I picked up that information on my own, without being conscious of it. Like photographing the Cosmic Enersphere in Moca, my little Minolta was aimed at the right unseen energy focus.

If NASA is interested in Santa Montana, perhaps it is because of all of the activity around the crucifix. I have been told that the church's original book about Elenita was removed from the book store and has been revised by the Catholic Church. What could be that important or make that much difference, that facts must now be changed? Whenever questions about Elenita are addressed to the nuns and clergy that serve the chapel on Santa Montana, they are now reluctant to respond and appear nervous. They say that Elenita is no longer a person considered important or spiritually sanctified. The change in the Church's attitude about Elenita coincided with NASA's interest in the Santa Montana site. Why does NASA close the facilities for their own use? Perhaps Elenita was a space woman on a mission, and not a saint! To me, *that* is a real possibility!

VIRGIN MARY
MOTHER OF JESUS,
DAUGHTER OF THE UNIVERSE

I have become very much interested in the appearances of the Virgin Mary, all over the planet. Throughout the history of Christianity, the Blessed Mother has made her impact on the devout and the non-practicing alike. In all cases, Mary's energy has been reported as loving and powerfully exquisite. It also has been reported that she has manifested appearances before non-Christians and has shared spiritual information with these non believers.

In the last ten years, Marianist appearances have escalated world wide. So many people (called visionaries) have reported regular appearances of the Heavenly Mother in their presence that one must consider what is going on here on Earth! After hearing about Elenita in San Lorenzo, Puerto Rico, and formulating my opinion about her, I now have a similar opinion about all of these appearances of Mary. I strongly think that many of Mary's appearances are holographic images created by space or inter-dimensional beings. I believe that I now have photographic evidence to support my position.

The archetype of mother is sacred the world over. Mother imagery has been revered since day one in the history of humanity. Who doesn't love his/her mother or fervently desire love from a mother or mother figure! The essence of all that is female is the Goddess energy. In antediluvian times and the pre-Christian eras, the Goddess energy reigned supreme.

Currently, there is a resurgence of the Goddess. Women are banding together in groups to redefine and reawake the Goddess within and revel in it. It seems to be happening spontaneously. Therefore, I think it is a logical progression to assume that the most well known and most widely accepted mother figure/Goddess ever to appear in the consciousness of this planet, Mary, would take ascendancy in our time of greatest need and change. For non-Christians, the image of

Jesus creates a kneejerk reaction that is not positive. Much evil has been done to non believers in the name of Jesus that would cause their future generations to revile his name and recoil at his image.

It is not the same with the Blessed Mother. The reaction is not so intense. Her mantle is soft and maternal. She is much more acceptable to appear before people in a supernatural way.

Those space visitors who have studied us and continue to do so are no dummies. They know what we respond to and have studied our psyches very well. What image could deliver the message of peace, love and devotion to the Creator with the most sincerity, believability and acceptance? And boy, do we need to hear what she has been saying! The real question is, will we listen?

All of the Marianist information has been the same. We are in imminent danger of world wide catastrophe, if we do not get our act together and learn to love each other, and to put down our swords and turn them into plow shares. It is not a new message.

We have weapons of mass destruction, and the possibility of their use may be proliferating. Even with the downfall of Communism, these weapons are being made available to unstable insurgents and revolutionaries. Planetary love is losing to divisive and partisan politics. We had the opportunity for planetary peace with the downfall of the Soviet system. The gauntlet of war was picked up instead.

In the years prior to the cataclysmic warfare and genocide in what was Yugoslavia, the Blessed Mother appeared to young visionaries in Medjugorje. She foretold the future for that country and urged the visionaries to ask their countrymen to work towards brotherhood. Millions of pilgrims have visited the site and many miracles were said to have occurred there. The greatest miracle of peace and brotherhood has not yet manifested, however.

It is interesting to note that sightings of UFOs have often been seen during the apparitions. When the Medjugorje visionaries asked the Blessed Mother about the UFOs, it is reported that she smiled and did not respond with an answer.

In South Florida, Nora, a visionary, reported having regular appearances of Mary on the 13th of each month. The mechanics of the manifestations of Mary are usually the same. For a short time, or for several years, she appears to certain visionaries at regular intervals. There are certain physical phenomena that usually happen. Most often it is the pulsations of the sun that are the most dramatic. The sun appears to vibrate and move forward and back, as if it has been zapped from its normal position. It is most often, at this time, that the Blessed Mother appears to the visionary(ies).

I visited Nora's home on the 13th of the month and took some very interesting photos, quite unexpectedly. At the time we were alerted that the Virgin was arriving, I aimed my simple camera skyward and was very surprised at what developed on the film. In 4 of the photos, there was a hard edged, Saturn shaped craft and a circular object with it that appeared to have a laser type device on it. I believe that it was laser, since it emitted a thin, rainbow colored portion of light, like a beam. In one of the photos, there was what looked like an apparition of the Virgin. It resembled her pose, stance and flowing mantle and cloak. She had her hands clasped in the prayer position and her head cocked to the side. It was a typical Marianist pose. In some of the photos, the Saturn shaped craft is positioned behind the clouds, therefore eliminating a lens flare possibility.

Thousands of visitors came to Nora's home until the Roman Catholic Church forbade Catholics from visiting or validating the apparitions. Despite that, many people said that they had miracles occur there. One month, an incredible circular rainbow was seen exactly at the appointed time of the Holy Mother's appearance. Nonetheless, the Church took a vehement posture of denial of the visions and the visionary. Perhaps, someone else photographed the Saturn shaped craft, as well.

Regardless of one's belief or non-belief in Christianity, the Roman Catholic Church or the Virgin Mary, the increasing numbers of what may or may not be Marianist apparitions is profoundly important. Her message is truly universal and needed. It is not a sectarian point of view. We are being warned that we must change our ways and learn to come together as one race, the human race, or we will no longer have a livable planet. Using our free will, we have an opportunity to change our destiny.

Other universal species are very concerned about what we do here. If we destroy this planet, it will impact upon this universe and the multi universes. Energy is never lost, it can only be transformed. We are a drop in the ocean of consciousness and the ripple is felt everywhere in God's creations. We are not separate, all is One. Perhaps if we are not listening to the lessons being forced upon us gently, it will be hammered on us from beings who are not human.

Someone is trying to get our attention! When are we going to take heed, before it is too late? No matter what our belief systems, we should listen to the Blessed Mother and her spiritual wisdom. We must unite in the love of the Creator for the good of All and do it now!

Above and below: Photos taken in Hollywood, Florida at the home of Nora, the visionary who reported having regular appearances of Mary on the 13th of each month. The "ships" and figure of the Virgin were not visible to the naked eye. ©1995

295

There are seven major energy vortexes in the world and Shasta is one that extends out one hundred miles in circumference from the center of the mountain. Within this circumference are sites of accelerated energy.

MT. SHASTA-
DOORWAY TO THE UNSEEN WORLDS

Those who are metaphysically based and interested in the unseen realms in fact and legend should make it a point to visit Mt. Shasta in north central California. It is a powerful and challenging experience. Your spiritual battery will be charged up just by the sheer beauty of the area. Visiting the mountain had been on my mind since I first heard about it. I could not finish my Earthly existence without a pilgrimage to the area. It was an absolute "must" and needed to be experienced.

The most difficult part of the journey for me was just being in California. As incredibly beautiful and dynamic as the state is, whenever I go there I am very uneasy. I feel like I am standing on a powder keg and the plunger is ready to be pushed. So the press of going to Shasta was even more urgent in my mind and heart. I have a feeling that it soon will be a very different place geophysically. I had no idea that Mt. Shasta is a volcano that is currently dormant, which I discovered when I went there. I always try not to be too well versed on the areas I visit that are vortexes, since I want to get my own feelings and reactions that are a priori. I am glad that I did because Shasta is a bundle of surprise packages.

It is no wonder that Mt. Shasta is a magnet, since there is so much mystery about it. It is said to be the home of the ancients, especially the Lemurians, whose descendants still are said to live within and beneath the mountain. Many visitors and locals have said that they have seen evidence of this in encounters with unusual beings and lights that glow from the slopes of the 12,000 foot mountain. There is always snow on the peak and often lenticular clouds, that look like space craft, blanket the mountain top. UFO reports are ubiquitous in the area and especially in the vicinity of the mountain. It is said that a tunnel system exists linking Mt. Shasta with the Andes Mountains in Peru and the Mato Grosso jungle in Brazil. Interestingly enough, a high speed tube transit system supposedly exists linking various subterranean areas within the Earth's crust. These Lemurian descendants are said to have mastered space travel and are members of a federation of planets. I had heard similar information when I was in Peru. The Peruvians talk about the spiritual White Brotherhood using the underground link between Peru and Mt. Shasta. These areas are supposed to be vortexes that are polar opposites, energetically.

My sister Suzy and I found much to do in the area. There are so many sacred sites to visit, each place is loaded with anecdotal information that is available about it. I have the feeling that you can spend a lifetime there and still discover new things about the area. Since we didn't have a lifetime, we decided to focus on eight sites. It was a worthwhile effort and quite a different experience than what I had in Sedona, Arizona. I thought, "How different could the Mt. Shasta area be from the red mountains of Sedona", but it was *very* different and very unique!

There are seven major energy vortexes in the world and Shasta is one that extends out one hundred miles in circumference from the center of the mountain. Within this circumference are sites of accelerated energy. This is not new found information, as those who have lived in the area, especially Native Americans, have regarded these sites as sacred ground. Still today, they are much revered places of ceremony. As a result, those from other spiritual realms, angelic beings, Ascended Masters and Saints are attracted to the area and have been seen trodding the paths.

As we entered each vortex area, Suzy and I prayed asking for permission, from the spirits that resided there, to enter their domain. Each was awe inspiring, in its own way. Richly verdant forested areas, craggy granite rock spires, water falls, gentle streams and ponds, mineral springs and pebble strewn deep cave grounds were all worth experiencing! Not all of the areas were relaxing and meditative. Some gave me the willies and made me feel on edge. But it was a good "on edge", like how a child feels on Halloween! This eerie feeling came from my experience at Pluto Caves. Native Americans say that it is an interdimensional doorway to the world of spirit and inner-Earth energies. Lava flows created tube-like caves that are vast, deep and very dark. Interdimensional beings and Lemurians have been seen coming out of the caves and in the environs. These caves were difficult to find and not on well-marked entry paths. I was amazed that I seemed to be drawn to or guided to the right path that led directly to the site.

Suzy and I had the uncomfortable feeling that we were being watched or that we were not welcome to spend too much time near the caves. I felt like we were being tolerated. I can still have a visceral reaction to the place. I noticed as we were walking away from the caves that the stones looked very similar to the coquina or coral stone found in Florida's coastal areas. When I showed locals the stone, I was told that at one time, the entire area had been underwater, perhaps the Pacific.

At the end of our trip, we were determined to visit McCloud Falls. It is a collection of three falls in three separate forested areas. Native American energies are strongly felt there. The falls are known for their ability to dispel emotional disturbance and create balance. They are each situated in a wilderness of pristine beauty. There was a road that connected the three falls. The lower falls we found quite easily. We thought it would be as easy to find the other two, but we were mistaken. With map and guidebook in hand, we drove for miles without success.

Frustrated and confused Suzy and I snapped at each other. It was the first time in a week that we had any negative words. We finally saw a sign indicating parking, so we headed in that direction. When we arrived at the parking area, there were no falls in sight, just a stream with a dock area from which to fish. Surrounding the dock were extremely tall trees. Suzy, who is a garden designer and is well versed in everything horticultural, said excitedly that the tall trees were Redwoods. She said that they must be very old because they are so tall.

We instinctively walked over to the trees and we each found a tree to touch. Both of us took the same stance, palms down directly on the bark. As we were standing there with our eyes shut feeling the life force of these giants of nature, I laughed to myself about all of those who condemn the ecologist "tree huggers." There is no way I could be anything but proud to be a tree hugger. To deny our connection with nature is to deny our connection with the Creator.

We never found the falls we were looking for but felt quite content with what we had experienced. It was the perfect ending for our Mt. Shasta experience.

Two weeks later, after I returned home, Troy and I were communicating with Mortin, an interdimensional guide. Mortin said, "Much transformation is taking place within you now. Take hold and use your powers. Everything is now open and available to you." "Yes, " I said, "I am feeling very....." and before I could get the word out, Mortin said "Clear."

"Exactly, I think Mt. Shasta had something to do with this. Is that right?" I asked. Mortin responded, "Yes, the clearing took place within you where the tall trees were. You felt their awesome presence at your soul's depth. The power of the trees provided the catalyst for the beginning of your visions on a conscious level. You will find self through self and then others will follow."

It had not occurred to me that the experience of seeing the trees and being there with them was so important. In retrospect, I now understand why Suzy and I had such a difficult time finding what we were looking for. If we had found the falls, we would not have ended up hugging the Redwoods and experiencing the area where they stood. It was not what we planned! Hindsight is the best sight. Again everything was in absolute and perfect order, even going too far down the wrong road!

Now, months past my Mt. Shasta visit, I am still energized and imbued with the legendary spirit of the Mt. Olympus of America.

SURREAL SEDONA

The first time I saw the glorious red rocks of Sedona, my boyfriend, Jack, and I were driving through town en route to the Grand Canyon. I had been hurt and agitated by his actions some 1500 miles before and was in deep upset. Our trip was not a happy one until I gazed at the unexpected splendor moving in front of and around me as the car sped forward. We were both stunned at what we were seeing and decided to stop at a roadside outdoor hot dog stand to eat and bask in the natural beauty surrounding us. I can remember how the sun felt, the deep turquoise of the sky, and the clarity of the air. No restaurant ambiance in the world could have surpassed this roadside eatery. In my mind I thought, "I don't know what this place is or what it has, but I am coming back here again!"

That trip was a few years before I knew anything about my cosmic connection, or was conscious of my need to visit planetary vortexes to charge up my spiritual battery. After that first visit, I became attuned to hearing Sedona being mentioned. I found that most metaphysical people were very conversant with Sedona and what it has to offer. Many had made pilgrimages to experience the vortexes and meditate in the tranquil environs of Oak Creek Canyon. Sedona has become the New Age center of America. I have made two more visits since I made the mental decision to return and experience just what Sedona is, and I have seen major growth of the town since my first time passing through.

Roger, an Excyle, asked me to accompany him to Sedona. We met at the Phoenix airport and made the couple hours' drive to Sedona. I greatly anticipated seeing those red rocks again. As we approached the area, the sun was at its zenith and the brilliance of the red stood out from the surrounding earthy tones. Roger was seeing this for the first time and shook his head and sighed, "Wow, this is incredible! "We found our motel and set out to get information on the town and its vortex areas. All of the metaphysical book stores have an abundance of information on the area. We made our way to all of the vortex areas and meditated there. We each had our own favorite places. Roger resonated to Bell Rock and I preferred Coffee Pot Mountain, one of the lesser visited high energy spots.

Many times we sensed "energies" around us. Since we both were interested in things extraterrestrial and interdimensional, we were hoping that we would see something paranormal. It was disappointing that we didn't see anything, but we were happy that we had come to Sedona, despite that undelivered expectation.

Roger surprised me with a gift, a week after we had returned home. He handed me a small photo album and told me to open it. As I was turning the pages, I got a happy shock. In several pictures taken in two of the vortexes, Bell Rock and Boynton Canyon, emerging from the sky above, and directed towards us, were beams of light or energy. These beams showed up in pictures of us, both together and alone. Roger said he checked with the photo developer and that they were not glitches or mistakes in the film or developing. Therefore they must be the "real thing." What is fascinating is that the angle is directed *towards* us, not on the side or somewhere else. Obviously, *we* were the focus of interest.

Roger and "the beam" in Sedona. ©1995

302

Looking at the photos reminded us of the visit we had in Sedona with my friend, Vera. I had not seen or been in contact with her for several years. When I heard that she lived in Sedona, I decided to look her up. Roger and I had a great chat with her. We were surprised to learn that we both had spiritual interests.

Vera was training to become a Spritualist minister. When I told her of my interest in UFOs, she told me that most people who lived in Sedona have seen UFOs or had some kind of an experience. She said that she had a very peculiar experience and was sure that it had something to do with UFOs. Her story truly is bizarre!

Vera was going to meet some friends for cocktail hour at a local restaurant. It was in the early fall and was still light out at 6:30 when she went to pick up her friends. One of her friends was already in the car when they went to the home of another. Just as Vera got out of the car and was walking towards her friend's home to get her, she was enveloped in a bright blue beam emerging from the sky. She felt the beam before she saw it. Suddenly she started to have tingling sensations all over her body. She thought that she might be having a stroke, or worse! Vera's friend who was in the car saw the beam and started screaming for her to look up and around her! As soon as she did, she saw the scintillating brilliance of the beam and started to panic. She felt faint but got the courage to run towards her friend's house. As she did, she feared that her friend might also be caught in the beam, so she ran back to the car. As she ran, the beam followed her movements.

Hearing the screams, her friend and neighbors came out to see what was going on. They all saw the bright blue beam emanating from some unseen source above. If there was a space craft, it was in an invisible mode, perhaps using a cloaking device, a la Star Trek. As Vera entered the car, the beam disappeared. As the beam disappeared, it left bit by bit, from the bottom up, until all of it was gone.

Vera and her friends were totally shocked. Planning on a pleasant drink after work was *not* what had manifested. They went from a totally mundane reality to something absolutely surreal! Vera told us that she felt the tingling sensations for a week. She hadn't a clue why the beam sought *her* out. She was not particularly interested in ufology but after that experience, she always kept her eyes looking upwards. Was it possible that Vera's beam experience and our beam photos were more validations from the universe? Maybe it was just a cosmic greeting for Roger and me?

CHAPTER FOUR

ADVENTURES ALONG THE
SPIRITUAL PATH

*Why would I be seeing
ghostly apparitions with togas
on a beach in Bimini?*

BIMINI DEJA VU

When my ex-husband, Lee, and I moved to a lovely waterfront home in the Ft. Lauderdale, Florida area, the first thing he did was to purchase a Grand Banks trawler. Boating and fishing became his reason for being. Along with that came cruising to better fishing grounds. The island of Bimini in the Bahamas was the key spot for fishing sportsmen. It was about 50 miles from our home to Bimini, a slow, leisurely cruise in a trawler with lots of potential undersea bounty en route. It was decided that Lee would sail there with our friend, John and I would drive to Miami and take the Chalk's sea plane and meet them in Bimini. My sister Jeanne and her husband, Ralph would arrive by plane and meet us the following day.

I had never been to Bimini before so this was all new terrain to me. When the sea plane arrived it dropped the passengers off on the banks of the North Island. Bimini, itself, is basically composed of the long narrow stretch of the North Island and the much smaller South Island with a channel running between, leading out to the Atlantic and some of the best fishing in the world.

Our first day out through the channel was perfect. The sky was a cloudless, deep azure blue and the sea as calm as a bird bath. I liked trawlers because they had full waist high railings that went around the boat making it easy to stand and gaze at the passing scene. That day, I was standing alone midship on the portside looking at the South Island passing us by. As we cruised, we passed a tiny sandy strip of beach that was quite narrow. At that time, in the winter of 1975, there was nothing else on the South Island but a small airstrip and this tiny beach strip. Instantaneously, I could feel the hairs raise on my arms and goose bumps on my neck. I started to feel a bit weird, and then in what seemed like my mind's eyes and in mind's ears, I saw and heard something even weirder than I felt! Along that tiny beach strip, I saw several diaphanous figures standing several feet apart from each other. They were dressed in what looked like togas and all were wailing plaintively. There were men, women and children in ghostly, transparent bodies. As we passed each form, they moved their arms out fully in front of them towards the boat. I had the feeling they were directing their gazes solely to me. This occurred in a matter of seconds, but I was sure of what I had experienced and I was shocked.

I hesitate to say that I had actually *seen* anything. It seemed more like a conscious hallucination. It just didn't make sense, though. Why would I be seeing ghostly apparitions with togas on a beach in Bimini? It didn't jibe...limbo maniacs, maybe, but not toga wearing cry babies! My curiosity was definitely piqued and I couldn't wait until we returned through that channel on our way back to the dock.

The fishing day ended successfully. We sailed back laden down with many varieties of delectable fish. This time, my sister, Jeanne was standing next to me by the railing. She was chattering about the day's events and the upcoming evening's. As we approached that now much anticipated stretch of beach, it seemed as if Jeanne was fading into the distance and I could only see her with my peripheral vision. Once again those toga wearing phantoms came into my view and they still were bawling away.

I was totally fascinated. Peripherally, I could see Jeanne looking at me, shrugging her shoulders and walking away. I was totally ignoring her, but it didn't seem as if I had a choice. Quickly, we once again passed the area and headed on into home port.

That evening, we all had dinner at the Complete Angler Restaurant and Bar. It was the main hangout in Bimini and was notorious because it had been frequented by Ernest Hemingway and other mavericks of the past. We had a fine seafood meal there and then Lee, Jeanne, Ralph and I made our way back to our cozy bunks on board.

About an hour later, John returned. "You should have stayed a bit longer," he said, "you missed all the action. Jacques Cousteau and his men came by and we had an interesting conversation." I said, wide eyed, "Jacques Cousteau? Why is he here? Is he looking for whales or sharks or something?" "No," John responded, "he and his men are exploring off the North Island. They are investigating an ancient civilization that supposedly was here. They have found pillars and what appears to be a central square type area under water. They are quite sure that there is a lot more around here!"

John's meeting with Jacques and his men began to make my experience with the phantoms more valid. I still wasn't sure, about what I saw, but I now knew that I, indeed, *had* seen something, although nobody else on board our boat seemed to have. Lee and John were going to take the boat back home and the rest of us would fly back to Miami the next day, so I didn't have an opportunity to go through the channel again.

I decided that my yoga and meditation teacher, Pat Parker, would be the best person to discuss my strange experience with. Not only is she an astute teacher of metaphysics, she also is an adept fisherwoman and knows the Bahamas well. I was correct, Pat certainly knew Bimini and all the attendant mystery and history of the island. She would guide me through my next experience.

I had lived in what was known as Atlantis approximately 26,000 years ago. I was a female whose parents were both extremely important to the people of those times. They were the Keepers of the Crystal.

ATLANTIS DEJA VU

I looked forward to my first yoga class after my return home from Bimini. I was extremely curious about what transpired there. What had I seen and why did I see it? (If, indeed, I really did!!)

When we came to the deep relaxation and meditation part of our class, I shared my Bimini experience with Pat and the students. Even in the dimly lit classroom, I could see a smile come over Pat's face as I recounted all the details. When I finished, she asked me if I had ever heard of the lost continent of Atlantis. I said that I never had. She went on to say that there is much information, metaphysical and literary that says there was a highly-evolved civilization that flowered and then destroyed itself. Many believe that Bimini is the site of this lost continent.

I was more confused than ever. "Why should I have experienced those apparitions there? Why me?" The smile once again appeared on Pat's face. "Well, maybe you are just a good antenna and you are picking up the vibrations of that time, like a psychic. Or perhaps, you, yourself, have lived during those times and are having an opening of those lost memory banks. Your being in the area has stimulated soul recognition of that experience. It is quite easy to find out if you have lived there by using regression techniques in a deep relaxed state. Would you like to try to see if you have a connection?" she asked.

I had an experience using regression with Pat once before. To begin to validate this experience in any way, one must believe in reincarnation since regressions reach into past life experiences. I thought I did, it made sense to me but I still wasn't completely sure. I was pretty new to all of this, having been in yoga only for a few months, but it definitely was titillating to ponder. My first regression had revealed that I lived in 16th century Holland, was an apprentice to a lithographer and drowned when I went fishing with my father.

It was a very realistic experience to me and I started shivering as I saw and felt myself going deeper into the sea. I also saw myself as a male which amazed me. I had no idea that you can be either sex in various lifetimes.

I decided to take Pat's suggestion and plumb the depths of my soul memory to see if I had any relationship to Bimini or Atlantis. I went into a deeply relaxed state which took a few minutes, and through a process of visualization, I was able to find the answer to my perplexing question.

I had lived in what was known as Atlantis approximately 26,000 years ago. I was a female whose parents were both extremely important to the people of those times. They were the Keepers of the Crystal. The Crystal appeared to me to be gigantic and was housed in a building that looked similar to our atomic power stations. I was the heir apparent to their position as Keeper. They were teaching me their knowledge. I saw myself as married to an extraterrestrial who had come with others from his planet to genetically encode the Earth with what he called "'reason." He was in charge of the diplomats that were coming from other cosmic systems to help in Earth's development or just for the usual look-see of space travellers. We had twins, a male and a female. When I discussed the children I said that they were perfect children in every way. As I said this, I did not feel parental love. I was very proud but detached, more like a relative than a parent.

It seemed that there was a war between two factions, the Emotionals and the Rationals. The Emotionals wanted to harness the powers of the crystal for war and weaponry. The Rationals were adamantly against this and therefore a civil war ensued. My husband was, of course, in charge of the Rational contingent. The next thing I viewed was myself, my children and their nurse in a cave. There was some kind of a cataclysmic event and we died in that cave. No other information came through.

I was totally amazed at how I could get this information. I was especially fascinated with the business of being married to an extraterrestrial. I had never cultivated an interest in or even thought in terms of E.T.s or UFOs. Why would I bring that in? This event occurred prior to the current generally pervasive interest in crystals. Pat was also bowled over with all of the specific information. She said that much of what I had described was, in fact, stated similarly in the literature about Atlantis. I say categorically here that I had *never* heard of or read anything about Atlantis, crystals or E.T.s in ancient times. Obviously, this information was coming from somewhere. It would be 15 years later, that I would be getting my first clue as to how accurate my regression was.

"ON ATLANTIS, THE EXCYLE, ELLEN WAS YOUR SISTER WHO TOOK CARE OF YOU. SHE WAS YOUR OLDER SISTER WHO WAS VERY STRONG AND PROVIDED SUPPORT THROUGH BAD TIMES AND IN YOUR TROUBLES. HER NAME WAS GHELIA THEN AND YOURS WAS HELIO. YOUR FATHER, KEHELDER, WAS CONSIDERED THE KING OF COMMERCE AND ALSO THE BOSS OF THE CRYSTAL REACTOR. GHELIA WAS A FEMALE WARRIOR IN CHARGE OF DEFENSE. SHE WAS SECOND IN COMMAND IN LEADERSHIP OF THE ARMY. THE SOUL WHO WAS KEHELDER IS VERY POWERFUL IN THIS LIFETIME AND COMES CLOSE TO YOU SOMETIMES. YOUR HUSBAND WHO ABANDONED YOU IS YOUR ACQUAINTANCE, EXCYLE GREGORY."

– Zarg

313

It seemed totally illogical to take me off of the bus to do such a thing. I knew that this was a kind of subtle harassment. It couldn't be anything else but a mind game.

EPIC EGYPTIAN EPOCH

To experience Egypt was also on my mental agenda, it was a desire but not a deep yearning. Somehow, I felt that there might be some emotional energy attached to my visiting the sites. I knew that I had to see Egypt but it was somewhere down the list, although I wasn't as reluctant to go there as I was to visit Israel. I fear visiting Israel until there is a firmly established peace in the Middle East. Even though millions of people have enjoyed safe travel despite the continual guerrilla and open warfare, I just cannot go there until peace is a commitment by all sides.

Eventually, an appropriate and exciting trip to Egypt was offered by a New Age travel company. It was combining the Egyptian experience with an UFO conference with many notable UFO researchers. I was especially excited to have Zachariah Sitchen along. Sitchen is one the most eminent Sumerian scholars and advocate of the influence of ancient astronauts on Earth's cultural and genetic development. He also is deeply involved in researching Egyptian heritage. It sounded like the perfect way to see Egypt and check it off my list of power places to visit.

The group travelling came together in an airport lounge in JFK Airport in New York. The group came from all over the States. I recognized some UFO researchers waiting in the lounge, including Wendelle Stevens. The trip to Cairo was long but pleasant. Now, I was beginning to get revved up for the visit there. When we arrived at Cairo airport, all went quite smoothly and in an organized fashion. We were told to point out our luggage to authorized baggage handlers, after we cleared customs, and then proceed to a waiting bus that would take us to the famous Mena Oberoi Hotel, in view of the Pyramids of Giza. I boarded the bus with the others.

We were in the process of getting an orientation speech from a tour organizer, when an Egyptian official accompanied by two armed soldiers or police, ran up to the bus. The bus driver was just putting the bus in gear, when he was waved to halt. The official said something to the tour guy and then came onto the bus. "Mia Adams, please come with me," he said. I got off the bus and accompanied by the two armed police types, the Egyptian official and the tour organizer walked back into the airport baggage area. It was strange, but I wasn't fearful.

I had expected something to happen. The day before I left Ft. Lauderdale, I typed a list of names, addresses and phone numbers of my UFO researcher friends, my mother and my attorney. This was a list, not to send postcards to, but for an emergency, just in case I was put in jail for a trumped up reason. So, it looked like my hunch was beginning to pan out, I just hoped that it would end there and not lead to an escalation. I felt that whatever would happen would relate to the intelligence agents' surveillance and UFO research issues. I knew that the long arms of these agents reached everywhere on Earth.

In the baggage area, placed on a small platform, was my large suitcase. "Open it, please," the Egyptian official requested. I found my suitcase key and followed his directions. I opened the case and then turned around to face the official. The official standing several feet away from me and my suitcase said, "thank you, you may close it up now." The man did not even approach my suitcase, let alone pat it down or inspect it in any way! He kept a distance from the suitcase and me.

It seemed totally illogical to take me off of the bus to do such a thing. I knew that this was a kind of subtle harassment. It couldn't be anything else but a mind game. Since I was undergoing agencies' activities at home, they just wanted to let me know that they knew where I was going and what I was doing. Choosing to check me out, from all of the other travellers, including the V.I.P.'s, could only point in that direction! If the Egyptian official was looking for contraband or even just doing a cursory spot check, why didn't he look in my suitcase? It isn't paranoia to assume I was given a message, it is just clear thinking. What else could it be? What a unique way to begin this trip!

It felt like I was walking through a National Geographic article! To see the Pyramids of Giza and the Sphinx right out of your hotel window was thrilling. Interspersed with our UFO conference were trips to local places of interest. The high point of the Cairo/Giza stay for me was the three-hour meditation inside the Cheops Pyramid's King's Chamber. We were able to be there at night, when the other tourists were not allowed. We chanted and heard our voices resonate deep within the pyramid's interior. We each took turns laying inside the empty and open sarcophagus, which never held any Pharaoh's body but was representational of the eternality of the soul. We had more or less free reign to venture into the Queen's Chamber and the Pit. Climbing the ladder into the chamber areas is a real experience. I can't fathom how it is done with thousands visiting daily. I was glad that I wasn't a claustrophobe. I doubt if they could handle entering the pyramid. Just being there within the edifice at night was a great and powerful experience. I was surprised that I felt no deja vu. I guessed that I had no Egyptian connection in a past life, but I was wrong.

After we left the Cairo area, we cruised up the Nile to Philae in the South. It was an information packed tour to nine different temple and tomb areas. It was also a pure assault on the senses. There was so much to see and assimilate of monumental civilizations that are no more but have left reminders of their greatness and colossal proportions that are unrivaled in time.

I felt no personal connection of intensity to any of the sites until we visited Philae. We left the ship before 6 A.M. to arrive there by sunrise, when there were no other tourists around. Philae is on a tiny island in the Nile. We had to take a small launch to get there. The granitic rocks, columns and pillars created a spectacular vision, as we approached. The sky was turning a burnished purple, when we were welcomed to Philae by the night watchmen, surprised to see such early morning tourists. The air was cool and soft; as the sun rose, it sparkled off the sides of the columned walls like diamonds. Each of us went off to find a place to greet the dawn. I thought that at that moment, there was no better place to be. I had a tremendous sense of well being. I walked towards the main open courtyard. The high walls of the temple were carved with huge figures of gods. As I looked at them, I had a feeling of familiarity. My back began to stiffen, as my spine became ramrod straight. Suddenly, I had an impulse to put my hands together in a prayer position and walk through the courtyard.

The hairs rose on the back of my neck and I had an intense feeling of deja vu. Since our guide hadn't yet arrived, I wasn't sure of the purpose of this place but I had the knowing that I had, indeed, been there before. Suddenly in my mind's eye, I could see myself dressed in a flowing gown with a kind of veil, walking with other similarly dressed females, two by two across the courtyard.

I sat in the courtyard meditating, until the guide arrived and we were told to gather together to hear about Philae. We were told that this temple was dedicated to the goddess Isis, the goddess of fertility, the sister and wife of Osiris and the mother of Horus. As soon as I heard that, I perceived my experience in the courtyard was valid. I felt that I was a priestess in the Isis cult; I became aware of the devotion for the Isis ideal from that incarnation. It seemed so real in that environment. I had never given Isis a thought before!

On our way back to Cairo, the ship stopped at the Temple of Dendera. A couple in our group were going to have a wedding ceremony performed as a symbolic gesture of their spiritual union. I had a similar but less intense feeling there. Dendera is also dedicated to a goddess, Hathor and was also built in the Ptolemaic dynasty, as was Phylae. Perhaps, since these are contemporaneous structures, dedicated to goddesses, I probably worshiped at both. In both temples, I felt at home but not especially happy. So, if I existed in those times and served the goddesses in those temples, I was doing what was correct, but I was not blissed out. I feel that I am a lot happier in this lifetime then I was then.

Currently, I have learned to be the goddess in my own life. I no longer ask for results in supplicating prayers to the diety; I affirm that my desires are here right now and are mine by divine right of consciousness. Probably if I had known this cosmic spiritual corollary in my Egyptian days, neither Isis or Hathor ever would have said no to me! Maybe that is what spiritual development is all about!

DEJA VU-EGYPT

It is 2300 BC, my name is Clovise. I am an apprentice priestess at the Temples of Dendera and Philae. I have a cousin Ra-Ti, who is studying at the Temple of Papyrus, school for scribes. (In this lifetime, Ra-Ti is Troy. Ra-Ti is learning hieroglyphics. He wants to teach me to be a scribe, although women do not do that. Ra-Ti and I are half-Nubian and grew up together near Luxor. We always loved each other. Ra-Ti would spy on me when I would bring offerings to the altar in the center of the main courtyards, early in the morning. He tried to induce me to make love with him but I refused, although I loved him and wanted him also.

One day, his passions got too much for him and he followed me to a secluded area where I was praying and he surprised me At first, I thought he was just trying to scare me but then he raped me. This was all very forbidden stuff. I became pregnant and had to run away with him. We went into the most western part of Nubia and lost our identities there. We raised a family and I taught hieroglyphics to females. Ra-Ti had a scarab business.

– Mia
Self regression in meditation

319

*"I am the learning stone.
Remember that a stone is a
living thing and holds the
wisdom of the ages.
You must be as a stone
in your conviction.
Time is on your side.
It takes much energy to
wear a stone down.
Many a wave must
crush upon the stone
before it shows any wear,
even if you stand alone."*
– Learning Stone Spirit Guide

NEW MEXICO REDUX

In August 1989, I read a short article in the Ft. Lauderdale newspaper that not only intrigued me but in a way possessed me. The article was about Chaco Canyon, New Mexico. The canyon is on an Indian reservation near the town of Nageezi in north central New Mexico. The article talked about how magical and mysterious the canyon was and about the Anasazi Indians who had disappeared from the site leaving their energies behind. As I read the article, I knew that this was a place that *I had to go*!! There was no doubt in my mind that my journey there would be inevitable. It was as if the canyon would create a link that was waiting to be formed in the chain of my life experience. I did not recall the name of Chaco Canyon in my conscious memory, but it must have culled up something from my unconscious mind that needed to be completed.

I kept the article on my desk, glancing at it quite often, keeping the name Chaco Canyon alive on my mental agenda. I had other places and power sites to visit. Sedona, Peru, Stonehenge, Avebury, Glastonbury, Puerto Rico, Egypt and many towns where UFO conferences were being held. I had my priorities already strategized. I know Chaco Canyon would have its time in my life. In July, 1991, while I was in Sedona with my sister Suzy she said that she would like to go to New Mexico and especially to Santa Fe, where an astrologer friend of hers resided. Suzy and I share a common interest in metaphysics and she is open and receptive to my passion for ufology. We put the wheels of the Universe into motion and our trip to New Mexico for July, 1992 became a reality. Coinciding with the trip, MUFON, the largest UFO organization, was holding its annual meeting in Albuquerque the same week we planned to go. I was sure that all things worked for good for us.

As Suzy lives in the Chicago area, we met at the Albuquerque airport, rented a new Buick Regal and began our New Mexico odyssey. Coming from metropolitan South Florida, where wide open spaces on local roads are non-existent, I really loved driving on New Mexico terrain. Aside from the few cities with a bit of traffic like Albuquerque and Gallup, driving is pretty wide open. I took the wheel and we drove directly to the beautiful Spanish colonial town of Santa Fe. Before I left Florida, I went to AAA and had a trip tick completed. I was not concerned about getting to Santa Fe, Taos or any of the other touristy sites. My concern was solely getting to Chaco Canyon. It is pretty remote. The closest paved road outside of the canyon is 20 miles away. Dirt roads, often in poor condition due to rain and other inclement conditions, can be impassable. There were no services or homes on that route. Despite many cautionary warnings, I knew that with due care, we would not have any problems.

When Suzy and I discussed our ideas for the trip, she said one of her goals was to go on a "shamanic journey." I was just vaguely aware of what that meant. She explained that a person who was practiced in the skills of a shaman, or a Native American holy man or woman, leads you mentally through a process that provides personal insights and often cleansings and healings. The methodology includes beating of drums, chanting, incense from Indian smudge sticks, music and the vocalization of high pitched tones. Guided meditation or visualization is the focus of the journey. How to find an appropriate person to do this was Suzy's problem. We were not going to stay in any one place too long to find someone. I knew that whatever was best would demonstrate perfectly, and as the fates had it, it did!

After a few lovely days in Santa Fe, we left for Taos. I had been in Taos about eight years before and since then it had developed into quite a nice town, although it still was quite rustic compared to Santa Fe. We had a lot of fun shopping in Taos. It is a shopper's Mecca! Even the sales people are happy and very friendly. Suzy was exhausted and decided to rest. She sat down on a bench in a charming shopping plaza. I still had energy to spare and took off window shopping. I spotted a metaphysical book store and walked to the door. It was closed for the day but I noticed a bulletin board outside the door that was loaded with ads and notices for local metaphysical happenings and services. I thought that might be the place for Suzy to find the shaman for her shamanic journey. I ran back to her and told her what I found and encouraged her to check out the board. She gathered her packages and made her way to the store and the loaded board. Without any trouble, she found the name and phone number of a female shaman who seemed to fulfill the credentials needed.

When we returned to the motel, Suzy phoned the woman. Marika Bright is originally from Central Florida and had moved to New Mexico 15 years before. She was under the tutelage of a prominent local Native American shaman who had taught her the skills necessary for her to ply her craft. Mystical things came easy for Marika who had since childhood manifested psychical abilities. Marika is married to an artist and they make their home outside of Taos. Marika said that she was available to do a journey the next morning at 10 A.M. She said that she would come to our motel with her apprentice and together they would guide the journey. The journey process would take at least one hour to complete.

Suzy thought that it would be preferable if the journey could be performed at Marika's home, instead of a motel, but Marika insisted that she wanted to come to the motel. She said that the citizens of Taos were accustomed to the unusual nature of Native American traditions and ceremonies so we shouldn't be concerned at what the motel maids would think, hearing drums beating or smelling smudge sticks. Suzy agreed to the appointed meeting and then we discussed the logistics of the situation. Even though it was *not* part of my plan to go on a shamanic journey, it would be another experience for me on my path. Being in the midst of so much Native American energy, what could be a better place to do it?

We had to be out of the motel by 11 A.M. Our room had been one of the few in that place that was reserved, because it was a non-smoking room. The management wanted us out as close to the time as possible. Since we would only have one hour to complete the journey, from 10:00 to 11:00 AM, we would have to do it together. Suzy said that she had never considered doing a journey *with* someone, even her own sister. She always thought about doing it alone. I told her that was not a problem for me. I would be happy to bow out and just look at the mountains, or read a book in the lobby, while she was going on the journey. Truly, it was not important to me. Suzy said that she would sleep on the decision until the next morning and would then decide. I told her that either way, I would be happy.

Suzy's decision the next day was to have me journey with her. She felt that since we had come to New Mexico together for the experience, we should go the distance and do the journey together. It was meant to be! She was ever so right about that. It was the turning point for the trip for both of us.

Marika and her apprentice, Stella, came to the room right on time. They were loaded down with the accoutrements of their craft, drums, sticks, tape deck, crystals, feathers. It was interesting but I felt a little apprehensive. I couldn't help thinking about the people walking passed our door hearing and smelling strange stuff. Marika again reassured us that nobody would think twice about this. So, I decided to let myself get into the experience and forget the "real world" outside our motel door.

Suzy and I laid horizontally across our beds, each wrapped separately in a blanket. I closed my eyes and relaxed. At first, I felt very cold, as if I had been put into a refrigerator, I started to shiver but as quickly as it came, the cold left me. I then became quite warm and again cooled down to a normal feeling. If I was imagining the temperature change, it was pretty realistic! Then the process ensued, the drum beating, chanting, music that I knew, recorded at Machu Pichu, smoke from smudge sticks. I could feel the hands of Marika and Stella moving several feet above my body. Marika then spoke using creative visualization techniques. I was so relaxed, I was only half listening and cannot really remember what she said. I felt completely content. It was kind of the warm, fuzzy feeling one gets when he is happily sleepy. Despite this, my mind was still in touch with reality. I figured that I was at least in an alpha state, mind wise.

Seemingly, out of nowhere, out of my reverie, I felt a sharp pain in my heart area. It felt as if I had been stabbed! My thoughts quickly snapped to attention. What could that have been? Amazingly, instantly I got my answer. It is a cliche to say that one look is worth a thousand words but it is also true to this situation. The main difference is that the answer to my mental question appeared instantly as a drama played out for me, apparently in my mind! How this manifested I cannot say for sure. The answer probably was there all the time but it took the proper context, the shamanic journey, as a catalyst for it to be known. Nonetheless, I asked the mental question about what the sharp pain was to my heart and this is the answer that unfolded...

I heard a mental voice, mine or other, I don't know, say, "You were called Skybird in the canyon." I then "saw" myself in a female body. I was a young Indian maiden, about 14 years old. A brave about my age, who filled me with love when he looked at me, was trying to scare me by balancing himself on one foot near the precipice of a cliff. I was terrified but he continued until he lost his footing and fell into a ravine below. There was a shallow stream there. When he hit the stream, he broke his neck and drowned in the shallow stream.

As I, Skybird, saw what had taken place, I felt so broken-hearted that I made a decision that romantic love was no longer meant for me and that I would dedicate myself to the Great Spirit. The "voice" said that I then changed my name to Blue Claw. I saw myself, Blue Claw, as an old squaw. I lived apart from the pueblo and had become a shaman. My claim to fame as this particular shaman was that I created a very special concoction that served as both a panacea for illness and a brilliant blue dye for leather, especially. No other person had this ability. The ingredients were a combination of a certain rock and a berry pulverized together. The berry grew high above a special treelike bush. In order to get the berries, I, Blue Claw, had to shape-shift into a giant bear. (Apparently, shape-shifting is an ability commonly attributed to gifted shamen who can change themselves physically into animals at will. They also can return to the human form at will.)

One day, as Blue Claw had shape-shifted into a bear and was standing on hind legs while reaching for these special berries, a brave from the local pueblo saw the bear. Not knowing that the bear was really Blue Claw, he got very excited and thought what a great kill that would be for him. He could really make his mark as a great hunter with bagging that bear. He took his bow and arrow and shot the bear, Blue Claw, right in the heart! The bear was felled with one arrow. As soon as the bear fell to the earth, the bear shape-shifted back to the human form of Blue Claw. When the brave saw who the bear really was, he got terrified and ran away. He was so frightened at the prospect of killing the famous shaman, Blue Claw, that he never told anyone. The people of the pueblo never knew who or why Blue Claw was killed.

This dramatic scenario was offered or culled up from my unconscious as a reason for the shot of pain in my heart area. I thought that the story was pretty interesting and probably explained the driving force pulling me to Chaco Canyon but it definitely was NOT any thought that had occurred to me before. If I had thought that in a past life I had been an Indian, it was only in a very generic way, certainly nothing as specific as that life experience.

The shamanic journey proceeded for a half hour more past the Skybird/Blue Claw life experience and I continued to be in a pleasant meditational reverie. Then Marika proceeded to bring Suzy and me out of the journey. When we were sufficiently aware and awake, the first thing she said was, "Mia, Stella and I were doing a cleansing on your heart center. There was a lot of negative energy arising from that point. You must have had some problems in that area." When I heard her say that I was a bit startled. I quickly told her that I was not surprised and then proceeded to recount my feeling of pain and the mental drama from a past lifetime that had ensued.

When I finished, she and Stella had broad smiles on their faces. The synchronicity made them happy. Not only did that validate their expertise in their shamanship, but it maximized facilitating the healing. The acknowledgment of the subject to the experience is always the best salve to the wounds. Marika then proceeded to make a very important statement, although at the time, I didn't treat it with much more than a shoulder shrug. She said that while we were in New Mexico, we should expect and be aware of validations of the messages from the shamanic journeys as they were inevitable! Little did I know, how true that would be! Suzy had a good experience with Marika facilitating her journey but it was not a kindred experience to mine. Whether Suzy had been with me in my lifetime in the canyon, I do not know. But she was a party to the magical event to come.

The rest of the day following the shamanic journey, Suzy and I visited the Taos pueblo, headed south towards Albuquerque, and then proceeded west towards Chaco Canyon. We were truly loving our time in New Mexico. The people seemed to be the friendliest I have ever experienced anywhere I have travelled. Suzy concurred with that heartily. We stopped at a motel in a town approximately two hours from the canyon. We went to sleep early to prepare ourselves for the next day when I would finally realize my goal of experiencing the energies of Chaco Canyon.

We decided to go to a gas station to make sure that all was functioning well with our new Buick. It had not given us any problems, but I was still a bit concerned that we would be driving on dirt roads and be out of reach of any services. We did the normal routine of tire checking, gas and water filling and window cleaning and then proceeded back to the interstate to look for the road indicated by the AAA trip tick. I was driving and Suzy was navigating. As soon as we got on the interstate, a sign appeared indicating the next exit for Chaco Canyon. This was not the same as the route AAA had chosen for us!

We quickly decided to get off at the next exit and, after asking some locals who fully agreed that this was the most direct way to the canyon, we drove on. It was a terrific way to get to the entrance to the dirt road into the canyon. As I have stated before, driving in New Mexico is great. We drove for an hour on very excellent paved roads without seeing a soul except for a couple of cars. It is hard not to speed with such open space ahead and no cops around. I put the pedal to the metal and drove an average of 75 mph! Suzy is a bird watcher and an animal freak in general and had her eyes peeled. There was not even a sign of dead animals in the road, nothing! It was as if we were the only people in the world!

As I was speeding towards a fork in the road, where I was to make a left turn, I saw a black speck in the distance. It looked like it was in the middle of the road. I did not slow down and the speck did not move. As we neared it, we could make out that it was a dog, a big black one at that. It was facing us and not moving an inch. I was being forced to slow down by his sentry-like position. As we approached him, I slowed down to 45 mph and the dog ambled off to the right side of the road and stood at the shoulder, watching us as we passed. He then moved back into the center of the road and stood there again watching us. I looked out of the rearview mirror and Suzy turned around and we both thought that this was kind of a weird dog standing out there watching us. Where could he possibly have come from? There didn't seem to be any buildings visible. He became a passing thought as we whizzed on our way to the reservation road leading us to the canyon.

When we finally were on the 20 miles of dirt road that lead to Chaco Canyon National Park, we found that it was not as bad as we had been warned. The weather had been dry and the pockets of dirt were not a problem. I found myself driving quite rapidly for the conditions, about 45 mph, without any deterrents. My anticipation was mounting, as we neared the ranger's station and the parking lot.

My legs were getting cramped from the continual two hour drive. We had not stopped since we left the gas station. We parked the car and got out. I told Suzy that I wanted to get my backpack from the trunk. She walked from the passenger side and met me at the trunk. What we both saw instantly when we got to the trunk was unforgettable! It was a sight that astounded us and left us both speechless for the moment! Above the back left light extending to the left side of the license plate, were what looked like two prints of a giant paw and claw! The claw was twice the size of the paw. The print appeared to be made of dust but looked as if it had been painted or drawn on, although it had a three dimensional, bas relief appearance that would have been difficult to effect. The claw looked like it had been been bent in a clawing position. The talon appeared to have four or five "fingers" with nails that protruded like flames from candles. The joints looked gnarled. Six inches of space separated the paw from the claw. The much smaller paw had four pads with a very thick impasto of dust. It was if it had been troweled on. These were definitely two separate forms quite distinct from each other! How did they get on the trunk of the car? They were *not* there when we left the gas station two hours before, and we had *not* stopped until we arrived in the canyon! The rest of the car, itself, was pretty much free from dust as I could still see my reflection clearly on the polished finish of the trunk. My image was looming over the prints...me and the prints. The paw and the claw and me! Could it be that Marika's words to us after the shamanic journey were coming true? Is it possible that through some miraculous manifestation of the fates or spirit, my past life as Skybird/Blue Claw was being validated? The paw could represent the shape-shifted Bear and the Claw symbolic of Blue Claw and her/my abilities. The validation had manifested physically in the third dimension, not just as a thought pattern or a memory from the recesses of my unconscious mind. I got my camera and shot a photo of the two designs. The photo is interesting in that it encompasses me as a shadowy image reflected on the trunk taking a photo of a more visible and identifiable paw and claw.

The past made more visible than the present. What is reality and what is illusion? Can we separate our past from our present? Life is truly a miracle made manifest. I couldn't wait to find a Native American Park Ranger. I didn't think an "Anglo" would be able to accept this manifestation or understand it as well as a Native American. Suzy and I got information on the canyon's roads and pueblos and set out into Chaco Canyon.

The tranquility and magnificence of the canyon did not let me down. The pueblos and their attendant kivas, or ceremonial chambers, were in incredible condition for the hundreds or maybe even thousands of years since their construction. Although the canyon, itself, was kind of anti-climactic after our discovery on the car trunk. Near one of the kiva areas, I found a Native American park ranger. I approached him and asked if he had heard about any strange happenings of a spiritual nature in the canyon area. He smiled knowingly and nodded. "They happen all the time. We hear a lot of different stories, especially from folks who stay at the campgrounds overnight. Many have heard disembodied voices talking and many have seen what look like Indians wearing feathers and the rest of the old get-up. Nothing we hear surprises us." When I told the ranger about the paw and the claw, he laughed and said, "Well, now I have another story to tell!"

By the time we finished touring the canyon, most of the paw and claw prints had disappeared, since they were composed of dust or dirt. The photo turned out very well and is as overwhelming in Kodak color as it was in reality. But I needed to know more.

Two weeks after I returned home, I was communicating with my space brother Zarg. He brought up the subject of New Mexico. He said, "You had a great time in New Mexico!" I responded, "Yes, did

330

you know about the paws prints and all of the stuff that happened?" Zarg replied, "Yes. They were put on your car by the past life spirits of the canyon. They did it quite extemporaneously. The big black dog in the road was a shape shifter. It was the guardian of the canyon. It was waiting for you. The paw and the claw also were symbolic. The spiritual sign of the claw is fire. That is why it looks like candlelight on the tips. It represents bravery, strength, wisdom and enlightenment. The paw represents sensitivity, youth and kindness; 'youth' as in newness, reborn in a spiritual sense."

Well, I guess Zarg provided me with the rest of the story. It was now obvious to me that my trip to New Mexico and all that it entailed was destined to be part of my spiritual path. The messages seemed to be coming to me loud and clear. I had to tie up the loose strings on my karmic experiences. I have been fortunate to be able to facilitate the completions. Perhaps this was part of the test as to how well I am able to listen to the inner and outer voices and wisdoms that dominate our Earthly and other-worldly realities and dominions. When the inner and the outer merge and the melding can be seen, felt and heard, you learn who you are. I am eternally grateful to know who I am in the Universal Scheme. I give much thanks to the Powers that Be.

"The paw and the claw." ©1995

"Hope is the power to store the thoughts that heighten life. Hope is also reality providing the kiss of worthiness. Home is harmony within you.
– Zarg

ALASKAN COINCIDENCE...OR NOT

In June of 1993, another one of my goals was attained. I went to Alaska with a group of fellow travelers from my metaphysical church. It was both a land and water voyage. I have been fascinated by Alaska for as long as I can remember. There is something special to me about American outposts outside of Mainland USA. I had read a lot about the rugged individualism that makes Alaska special. Not only are people attracted to Alaska by its sheer physical beauty and opportunities to make a lot of money but also by the possibilities to be your own person undaunted by the constraints of civilization. Then there are the native cultures of the Inuits, Aleuts and the Eskimos that adds to the mix. I found it all as delicious and palatable a concoction in reality as I did in my mind! I was so very much drawn to Alaska. I found being there to be totally exhilarating.

One of my goals of the trip was to see the Mendenhall Glacier and to spend time on it. I planned to take a helicopter to the Glacier from Juneau. It was something I had to do! Given my South Florida constitution and my sheer aversion to anything cold, this little jaunt was not in my usual line of thinking. My cabin mate, Elaine and I booked the trip together and made our way to the heliport. We boarded the helicopter with two married couples from Arkansas and were flown to the Mendenhall Glacier by the youngest pilot I have ever seen. Even though it was June, it was a thrilling flight over snow capped peaks and verdant valleys with the Mendenhall's vast river of ice in front of us. We landed at the base camp there and were met by two employees whose job was to greet us and to move the base camp each day. Glaciers are in flux. They move and have fissures that open in different areas. We were told that was one of the fascinations for the workers coming to the base camp each day.

Mendenhall is a constantly changing panorama. The young helicopter pilot told us that he would be leaving us at the glacier for a longer period than normal, since he had an emergency medic evacuation in a remote area. Instead of staying there with us, he would leave us for one hour and then return. I had mixed feelings seeing him leave; it was kind of scary, but exciting!

In that hour, we found out a lot about the two young base camp employees and the awesome ice that surrounded us. It was a good thing that it was so monumentally scenic because it was also extremely cold for Elaine and me. I was glad that I wore my ski parka and my thermal underwear...I sure needed them! Despite the cold, I felt an electric adrenaline rush. I felt as if I was being charged by a battery. I felt truly alive with energy! I felt as if this was one of those moments of transformation that I have so often read about!

Being on the Glacier was like a meditation. It was surreal, inspiring and deeply relaxing. I never felt its like before. After more than an hour, the helicopter returned to the base camp. The pilot related the story of his medic evacuation. The patient had a simple broken leg but lived in an area reachable only by boat or plane. Alaskans were family and needed each other.

After another eye-filling ride, we landed at the airfield and took a bus back to the ship. Because of our trip, we had missed lunch. The ship was holding a special meal for us. As we approached the gangway to the ship, I suddenly felt a strong desire to walk around Juneau. We had a bird's-eye view of the very scenic town, with homes set on the sides of hills and along the waterways. I told Elaine to go on board without me. I wanted to see more of the town. I told myself that I was not going to do my normal "shopaholic" thing. I was just going to look around, and if the desire to enter a shop arose, no purchases would be allowed! I had already reached my nick nack maximum.

Alaska is a shopper's Mecca. The Native Alaskan crafts were superlative and irresistible. For the most part, I was true to my own self-imposed limitations. Wandering the streets of Juneau was fun and interesting. Most of the fine shops were treasure troves. I was drawn into the Snowsong Gallery by some unique sculptures in the window. I decided to have a look.

As soon as I stepped into the shop, I spotted a table with small soapstone and bone sculptures. I couldn't believe my own eyes as I viewed a piece displayed with the others. I instantly knew why I had to come into town! There in front of me was a figure of a very unusual shape. One end was a very identifiable bear with two legs and paws flat on the table, but the lower part of his body was the top part of a man, holding a spear in his hands. It could only be symbolic of a shaman shape-shifting into a bear! In all of the shops I had visited on this Alaskan jaunt, I had not seen anything like this! Wow! Now I knew why I couldn't go back to the ship. I was being drawn to this sculpture. It was a linking of my Chaco Canyon experience of the year before with my yearning to see Alaska. I also felt that the electrifying, transformative feelings I had just experienced on the glacier were connected with my finding this symbolic artwork. I just knew it!

I asked the owner of the gallery, Tom Blake, about the sculpture. He said it was a very unusual piece and had never seen any work like that done by its Eskimo sculptor, Ronald Komok. He confirmed that it, indeed, was a Eskimo shaman shape-shifting into a bear! He showed me a biography on the sculptor and it included two more coincidences/synchronicities. The sculptor's mentor's name is John ADAMS and the Snowsong Gallery is located on ADAMS Street! The same as my last name!

Needless to say, I couldn't resist buying the sculpture. I felt that it was mine by right of consciousness. The Universe led me to it. I knew that somehow my spiritual path was linked through my life experiences, and many of these incredible experiences manifested through consciousness. I was listening to my inner guidance. There is power in Power Places for me.

Alaska and the Mendenhall Glacier catalyzed my internal and external power sources. The shape-shifting shaman sculpture is an out-picturing of my spiritual listening device. It is a palpable, third dimensional gift, one that fits perfectly into my spiritual reality and neatly into my belief system. It also fulfills the ego's need to have something tangible for "proof" to fit into the life puzzle. Was it coincidence, happenstance, or just one of those things? I think not!

"Shape-shifting Shaman" by Eskimo sculptor, Ronald Komok, who also works under his Eskimo name, Panniniak. The body of the sculpture is soapstone, and the face and hands of the shaman are carved from walrus jawbone. ©1995

MY PASSAGE TO INDIA

We all have mental pictures of India. Some are overly romanticized, like the story of the Taj Mahal, while others are sadly dramatized as a level of poverty and human degradation we care not to see. India is all and none of these; it does depend upon your point of view. I don't think I had any serious expectations. I have been to many third world nations and had for a time lived in New York City. Seeing desperately impoverished humans was not a new experience for me. I was not inured to it, as I see it as a reality that exists most everywhere. Poverty in New York is the most dramatic and sadly depressing. The contrast in the have and have-not cultures is much more blatantly profound and terrifying to behold. It is not something to be proud of or acceptable anywhere but in India, there seems to be a tradition of it. Much has been written about the slums of Calcutta, the Untouchables and the caste system. We seem to know more about the poverty of India than we do the glory.

Fortunately, I again found myself with a group of strangers who were very appropriate for me to travel with in India. Leading the group was Fred, a former monk from a yoga discipline. He was now married and a father but still an intensely spiritual being. Joining us later on was Brian, formerly in the NASA astronaut program. He severed his ties with the program, before he ever had a space mission. Although Brian was a scientist, he was also metaphysically oriented. He was a devotee of Sai Baba and he and Fred had audiences with Sai Baba in the past. It is not an easy thing to do, to gain entree to a personal tete-a-tete with the guru. Sai Baba has hundreds of thousands of devotees whose greatest dream is to be within an eye's view of him. With Brian's past relationship to Baba, we luckily and gratefully were granted an audience with him at his home ashram. There is much ceremony that is connected with Sai Baba's daily walks through his devotees. He is as reverentially loved and treated as if he was a living god. I am not one of his followers, nor am I a devotee of any human. I also will not judge those who are. I don't like to put humans on pedestals because I don't believe any human can stand up to the scrutiny. In a way, I felt kind of guilty about having the opportunity to see Baba, just because I knew the right person. Nevertheless, I was very happy to get to do it! When the appropriate time came, after the ceremony of Sai Baba's entry amongst his people, we were beckoned into his office for the audience, while thousands of his devotees silently waited outside until he returned to walk amongst them again.

I don't feel it is fair for me to give my opinion on the experience. My nature is egalitarian and, as impartial as I think I might be, I still am a skeptic about those who claim to have "superhuman" abilities, and who have thousands of followers who believe in them with blind faith. The only reason I am even discussing this experience is because of my serendipitous connection with Baba after I returned home, and my surprise about how I later felt about Mother Teresa.

Again, Fred and Brian proved to be well connected. They had, met with Mother Teresa before and said that we would be able to meet her and see her facilities. I really wanted to see what she was doing. She has had such good press about all of her efforts, and has for so long been considered a living saint, but my skeptical self had a hard time believing it could possibly be true. I had read about her dogmatic Roman Catholic attitudes related to birth control and other women's issues. Being a feminist, I really took umbrage at that! My back was already arched. If she was a saint, I was going to look for the proof. I didn't have to look hard or long. If the proof of the pudding is in the making, Mother is a pretty top notch cook!

Upon arrival at our hotel in Calcutta, we were met by Arnold. He is a big, burly guy with a broad smile and a twinkle in his eye. Arnold comes from Vancouver and had travelled with Fred's group two years before. He is a very successful and quite wealthy insurance executive who became totally smitten with Calcutta and Mother Teresa's facilities. After his initial trip there, he vowed that he would return the next year and spend a couple of months volunteering in the famous Dying Destitutes facility. He kept his vow and this was now his second year volunteering. He was a very happy camper! As a matter of fact, he seemed totally joyful, if not ecstatic. Arnold greeted us as if he knew us forever. He is that kind of a guy, very irreverent but loving. Calcutta is not an easy place to live. It is totally pollution-ridden, traffic-congested and seemingly unliveable. There are more people per square inch there than anywhere else in the world. At night it is a strange tableau in the streets. All activities–from living, eating, sleeping, entrepreneuring–is done on the sidewalks; all you can see is teeming humanity everywhere! It is absolutely Felliniesque, surrealistic.

There is nothing that Arnold likes more than the streets of Calcutta. He purposefully chose to live in a third rate hotel that the Indians would choose. He did not want to live in the fancier hotels that most of the foreign volunteers stay in. He loved going native. I saw him kid around with a woman cooking for her family in the gutter. She offered him some of what she was cooking and he ate it! He thought nothing of eating in the street, a decided no-no for tourists. He said that he never got sick, and that included drinking the water! He was fearless also in crossing the streets in the middle of the block. There are no real rules of the road in India. Anything and everything goes in driving. To be a pedestrian is a real risk but Arnold loved to play in the traffic!

The most remarkable thing, that is so enduring in my mind was his love of serving in Mother Teresa's facilities. Arnold said that when he was in Calcutta, that is when he felt the most alive! His wonderful, fruitful and comfortable life in Vancouver was only a "half life" compared to the time he spent in Calcutta. He was going to make arrangements for us to see the Dying Destitutes facility, where he worked most of the time. He also volunteered in an outpatient facility in one of the worst slums and liked the one-on-one with the locals. He towered over most of their bowed and puny physiques, but was a gentle giant to them.

339

Unfortunately, Mother Teresa could not make her appointment with us. She had been called to attend a charity drive in the United States and left right before we came. She did leave special medals that she had blessed. Mother may not have been there in the physical, but her presence was definitely felt. Everybody that I spoke to expressed glowing respect for her. It wasn't just her selflessness, it was her inspiration. The one question that she asks everybody is "What have you done for God today?" It might seem like a cliche, but it is quite profound in its implication. Walking into the Dying Destitutes Home is like walking into God's house. There is no place I have ever been that I have been more impressed and touched.

Most people would think that The Dying Destitutes Home would be a depressing and yucky place, but it is the antithesis! Arnold showed us what the work there really is. The basic concept of the place is that it is a place to die with dignity, knowing that you are loved and cared about. That might seem like no big deal to most of us who take that for granted, given our more comfortable circumstances of life. For the average destitute Indian, it is a very big deal! Thousands, if not millions of beggars and poverty stricken humans, many terminally ill, line the streets, sewers and gutters of Calcutta. No one has ever shown them any respect or love. It is the most difficult and most futile of existences imaginable.

Mother's goal was to give these beings a place to die with dignity, where they would be loved, nurtured and cared for. No one would be turned away. There are even teams of volunteers who go out to look for those who are dying in the cracks of a pitiful life structure. The main train station is one of the places the teams go to find those who are so ill they cannot bring themselves to the hospice. Often they find them lying in their own vomit and/or feces, too weak to move.

Once in the facility, they are bathed and have a cursory examination. A clean mattress and linen is given them. For most, just having a clean, safe place to lay their heads is like being in paradise. The one criticism leveled at Mother is that these "clients" are not being given heavy duty medical treatment to cure them. Mother maintains that is not the purpose of the facility. The purpose is for the dying to feel loved and cared for in preparation to leave the physical body and move to the spiritual level. Since most Indians are not Christian, there is no attempt at proselytizing. The only thing given is service to humanity. There is nothing antiseptic about the facility. It is clean and well organized, staffed with mostly foreign volunteers from all over the world, Catholic and non-Catholic, each imbued with the desire to serve with love.

The most seriously ill are placed closest to the door, where they are given the most attention. Other clients sit up and chat amicably with each other and the staff. We watched as Arnold stooped down to talk to an obviously seriously ill man. He was pitifully thin and fragile looking. His face was hollow and expressionless except for his eyes that gazed into Arnold's. Arnold put the man's head into his large palm and with his other hand, he gently stroked the man's head and brow. While he was doing this, Arnold spoke to him compassionately about the spiritual journey upon which the man would inevitably soon embark. The look in the man's eyes was unforgettable. The communication was not just man-to-man, it was soul-to-soul. They both looked truly content and peaceful. There was no love giver better than Arnold.

I noticed that despite the diseases that the clients had, no volunteer was wearing a mask or gloves. Most just wore an apron. I mentioned it to Johann, the young German in charge of volunteers. He said that since the facility opened in the 1950's, no volunteer has ever worn a mask or gloves, despite the often communicable diseases they were dealing with, like tuberculosis and now AIDS. Johann said that there was something special working there that protects them and keeps them safe that transcends disease. We remarked to Johann how impressed we were with the volunteers and their selfless devotion to their clients. He then said to us, "We get far more from them than we ever possibly could give!" He had been there for seven years and planned to stay for many more!

The facility is divided into separate areas for males and females. Both were equally impressive. We also visited the infant and children's hospital, where good works of a different kind are performed. Always there was the utmost concern for those being served. I feel sorry to have missed meeting the persona of Mother Teresa, but I did not miss meeting the spirit of Mother. She is represented by her good works and those who seek to emulate her tireless service to humanity. What she has become for this planet is a catalyst for good, an example to be followed, a human devoted not only to God but to the God in us all, I feel that I have met Mother Teresa, not because I have shook her hand but because I have seen her heart.

> *"To Grow you must attract your own types to nurture you. To find a love, you must learn to love and romance yourself and then you will be fulfilled. To love someone without loving yourself is like having only half a heart."*
>
> – Zarg

BERYL'S STORY

I met Beryl through my friend Lynn. Beryl lived in New York and was the editor for a United Nations newsletter for UN employees. She was also one of the founders of the Parapsychology Club at the UN, which eventually hosted many well known UFO researchers and scientists who advocated the extraterrestrial hypothesis. Beryl was a feisty personality who had travelled extensively in India and Nepal. If ever there was an Indiaophile, it was Beryl. She eventually married and divorced a man from India, which despite her feistiness and hard edge was both her greatest happiness and sorrow. She never got over him, even until her dying day. In every conversation I had with her, she brought up his name.

Soon after she and her husband returned to New York, kidney disease plagued her. She became extremely ill but refused, for reasons known only to her, to have a kidney transplant, which probably would have stemmed the tide of her illness. Instead she sought out conventional treatments for her disease, which included exhaustive daily dialysis. She also became a strict vegetarian. No doubt the downward turn in her disease's pathology and her physical deterioration contributed to the decimation of her marriage. Despite her angst at her lost love and her bitterness over that, she did make attempts at understanding her lot in life.

As a child, she had been given away by her poverty-stricken parents to a neighboring couple. They adopted and raised her. It was not the happiest of childhoods but Beryl sought to make the best of it. She wanted passionately to find someone who loved her to her inner depths. As life would have it, she was rejected by her greatest love and given away by her natural parents. It was no wonder that she had a tart tongue and a tough countenance. She was a survivor but was now holding very tenuously to a lifeline of health and went looking for answers to some very difficult questions.

Lynn thought that we should meet. She thought that perhaps Beryl needed some spiritual ballast to help her through what was now looking like her toughest hours. Lynn felt that I might have some way of communicating with Beryl, since I had not only my ET. relationship but also was a meditator and metaphysician. I was appreciative of her confidence in me and said that I would try. Lynn warned me of Beryl's appearance. Even though Beryl was our age, she looked much older. Due to the progression of the disease, Beryl was about four and a half feet tall, stooped and withered with a dowager's hump. She looked nothing like her photographs of a few years before. Despite that, her voice and energy were strong. That always amazed me.

She asked me the toughest of questions about the nature of reality and life and how to deal with her health. We had many discussions both during my visits to New York and on her visits to Florida. We also had a pen pal relationship. It was in those letters to me that Beryl asked the most difficult questions. It is always the most profound task to try to bring solace to those you love, when they know that their days on the planet are waning.

The following is a letter I wrote to her in response to her queries.

> *Your question was very difficult and very powerful. There is no one who can know your suffering or your yearnings like you do. For it is only in illness that we can appreciate what it is like to be a healthy person and the advantages of a painless and fearless life. So as I was thinking about what I could say to you, everything I thought of seemed facile or a cliché. But as trite as cliches go, there is always truth attached. So here goes my response to you, although, it does, indeed, seem inadequate in its simplicity. But, oh my dear, it is soooooo difficult to embody.*
>
> *First of all, I applaud your search. Luckily you have an opportunity to experience wisdom and New Age thinking via your parapsychology club. I am sure you know your disease*

better than even your doctors and what makes you feel better. But I have been "Told" to offer you a different approach. My ET's response to me when I last had a major bitching and griping attack about when my "tests" were going to end, said , "celebrate your tests!" Wow, I thought, how absolutely simple and powerful and TRUTHFUL a statement! If you believe in the concept of Karma, you know that it is what you do with what you've got that counts on your spiritual path. These "tests" are how we evolve spiritually. That probably is no solace to you but it is one explanation of adversity.

I assume that you have already approached all the healing modalities especially visualization and meditation techniques. Bernie Siegal and the Symingtons, famous writers on self healing are musts for your reading list. It is hard to begin the healing process when you are still imbued with fear. Fear keeps us grounded spiritually. You must unite with the oneness of the Universal Divine Intelligence. That is in knowing that we all are part of God. We come from the Intelligence of Creation and we all return in the cycle of life. When you affirm your health, you must know deep within your being that it is, indeed, accomplished. There should be no fear, just a feeling of joy for being here and knowing your connectedness to all.

Like the Samurai, you are a spiritual warrior. They were known for their strength of character, tranquility and detachment. In their acceptance of reincarnation, they rose above the fear of death. The attitude of detachment is attainable by all who seek. It becomes the way one experiences life. You asked if there is anything else you can do to achieve healing. I think one very important way is to dwell on what life you can live now and the joy you feel in experiencing and functioning whatever way you can. Experience now! Know that happiness and health is a now function.

None of us is ever in touch with our mortality. We all assume that this is a forever life. If we worried each day if we would die in the next instant, we would loose the aliveness of the NOW MOMENT. And the truth is that we never die. Our bodies are temporary but our soul is eternal. That is why it is so important for us to work on our soul development. That is far more important than the temporal state of our corporeal reality. The body is just a vehicle for housing the soul. We are a soul with a body, not vice versa! If you feel that you are tired of being sick, that is an important decision. In detachment from the illness you will find a right attitude that will provide a new bravery. In that bravery evolves universal love, benevolence and compassion. And, through that comes sincerity, devotion and loyalty to soul self and mankind. In that will come right action - whatever that may be.

I hope you can understand what I'm saying. I am not asking you to forget your illness and pretend that you are healthy. I am saying address each remaining moment of life gloriously. Enjoy it. Perhaps your job in this lifetime is to live in the moment with joy, verve and ecstasy, as if it is the last. I think that would be an excellent way for us all to look at life, it should be the most natural in the natural scheme of things. Any way else is man's distortion of his own reality.

I hope this has helped. For us all, this is a glorious time to be on this planet. We are moving into a new and higher vibration and that heralds miraculous changes as is evidenced in all the political changes that have happened relatively bloodlessly. If this can happen in our lifetime, have faith, anything is possible.

In the love and light of the Infinite Creative Intelligence, I AM,
Mia

Soon after my letter, Beryl was no longer well enough to work, even part time, at her job with the U.N. It saddened her very much to have to quit. She enjoyed the camaraderie and interaction with her friends there. It was the well spring of her life. Word spread amongst her friends. Beryl had also started a Vegetarian Club at the United Nations that shared information on holistic health and vegetarian recipes. Some of the members of the club were devotees of Sai Baba of India. Service to community and mankind is a mainstay of the guru's teachings. These devotees found out that Beryl was in need of the kind of care that was not provided by local governmental help.

Beryl had one brother who was not close, emotionally or physically. She had no family to help. These Sai Baba devotees, those who worked at the UN and others from the local ashram, Beryl affectionately called "Babalinks." They became true friends and family to her in her last days. Finally, it appeared that God had graced her with the kind of unconditional love she had yearned for all of her life. Not only were they loving and nurturing, they also gave her the emotional and spiritual sustenance she needed.

Beryl also became a Sai Baba devotee and studied Hindu, Buddhist and Taoist philosophies. This became some large measure of preparation for her true journey home, out of the body that had held her hostage for so many painful and life-sucking years. Beryl had a large apartment, the one good thing left over from her days with her ex-husband. One of the "Babalinks," at least, was there every day, one even lived on the premises. No longer was she ever alone, physically or spiritually.

Beryl's greatest hope and dream was that she would regain strength enough to visit Sai Baba's ashram in India. He had become her center of solace and guidance. In a way, he became the God in her life.

I was not aware of Beryl's total immersion through love into Sai Baba's influence and teachings. As she was becoming increasingly weak, she could no longer speak on the phone or even write letters. I had lost touch with her. A week before my own trip to India, Lynn phoned me to tell me of Beryl's death. In my heart I was happy for Beryl. I felt that she had suffered enough in this lifetime to warrant a happier life the next go around. Lynn said that she had been summoned to New York by Beryl via the "Baba-links." Even though Lynn has major responsibilities on her job and a teen ager at home, she decided to go. Her long time friendship with Beryl had its difficult moments but they had a strong enough bond that Beryl wanted her at her bedside. It was to be full and loving closure for both of them.

Lynn told me in detail the beauty of Beryl's final hours. All of the "Baba-links" and a Hindu priest helped and aided in her transition to the other side. Lynn held Beryl's bony and wracked body in her arms, as Beryl took her final breath. The one thing that Beryl regretted was that she was not able to make a pilgrimage to Sai Baba's home ashram in India. It had been her fondest dream, which now went unrequited.

I told Lynn that on my forthcoming trip to India, plans were made to stay at Sai Baba's ashram and, if we were lucky, have a personal interview with the guru, himself. Since I was not a devotee or had much interest in any human gurus, this was now going to take on more importance to me. I was going to be Beryl's surrogate. I would have Beryl's spirit with me and do what she was not physically able to do.

My experience at the ashram is discussed in the prior chapter. I bought an Indian cloth to lay on in the ashram dormitory. After our days there, it was permeated with the incense of the ashram. Upon my return, I gifted it to Lynn. She was touched by the synchronicity and the irony of the event of my going to Sai Baba's home and the physical gift that carried the energy of the experience. Lynn told me that the following week, she had been invited to attend a wedding in Texas of Beryl's roommate, the original "Baba-link" in Beryl's life. She said that she would take the incense laden cloth with her and share the my experience of Sai Baba with her new friend.

The wedding was a combination of Christian and Hindu rites. The bride wrapped the cloth I gave to Lynn around her in the fashion of an Indian sari. Lynn said that everyone there who knew Beryl touched the cloth, knowing that this inanimate object was symbolic of Beryl's hopes, dreams and transcendence.

Lynn now has the cloth as a wall hanging at home. Without my even planning on a conscious level to aid in Beryl's final hours, somehow the Fates made me an instrument of her completion on this plane of existence. That for me was another miracle and an added blessing to my life.

> *"An illness is a lesson.*
> *Mend your mind and*
> *your body will follow."*
> – Zarg

*"The hardest thing
in life to learn
is which bridge to cross
and which bridge to burn!"*
– Anonymous

AND IN CONCLUSION....

There is one thing I can say about my life with certainty and that is that it isn't boring! The majority of the powerful and amazing events in my life happened in the process of just living. I did not purposefully, at least not consciously, seek out bizarre results or effects. Yet, there have been so many unusual, dare I say extraordinary, circumstances I have often encountered in so many places including my own home that it was very perplexing to me. My family and friends have come to expect these experiences no matter what I do. If I have average days, we all are a little disappointed. In this book, I have tried to get a grip on what it is about me that attracts this stuff but it is not an easy task.

My favorite intelligence agent/"son"/victim Jordan said that I am not the average woman. Zarg, my favorite interdimensional being, said that, too. That sounds so sanctimonious I hate it. My hypothesis is that perhaps some part of my belief system that is so free of doubt, so deeply subjective and instinctive, it attracts results that outpicture paranormally. If I was to believe Zarg, he would say that this certitude derives from my Excelta lineage and the karmic sum total of past lives I have had. Jordan would say it is because I have special abilities due to governmental human/alien machinations. Perhaps if Jordan had not entered my life, my questions about myself would have been a whole lot easier to answer. I could have just had a nice fuzzy metaphysical feeling that some beings way up and out there know me and love me and can't wait to have me back with them. It would have been nice and personal, just me and that part of the cosmic plan. A bigger, more convoluted question and less spiritually redeeming question is how and why the United State Government got involved in my life since childhood? If Jordan's information about this is a ruse, why bother? I really don't think I am that important to disinform, or am I? Maybe I will never know.

Jordan is still an active FBI Special Agent, working at the same office as when we first met. If there were any complications from our relationship or from his travels to Nellis Air Force Base and Dulce, New Mexico, it doesn't seem apparent. He has made good his promise to not contact me anymore. Six months after our last conversation, I decided to leave a message on Jordan's voice mail at the FBI office. The condomaniac, Mark, was going berserk again and I wanted to keep Jordan apprised of the situation. I did not expect that Jordan would answer his phone, but he did and I was not prepared for a personal conversation. I gave him the salient details of the current Mark situation and he again said that there wasn't anything they could do unless Mark did something illegal. After that statement, he didn't say anything. There was the proverbial pregnant pause and I asked him if he was okay and he replied in the affirmative. When he didn't respond further, I said that I was sorry to have bothered him and then hung up.

Either Jordan was shocked at my call, warned not to speak to me or didn't give a damn about me anymore or ever. What he did say was in a very detached, business like tone, totally unlike the Jordie I knew. No matter what his reason, he has severed our relationship.

So I am back at square one, still trying to figure things out. Maybe one day, like Jordie said, someone will tell me what the real truth is about me. But in the meantime, I won't hold my breath!